The Order of Time

ABOUT THE AUTHOR

Carlo Rovelli is a theoretical physicist who has made significant contributions to the physics of space and time. He has worked in Italy and the US, and is currently directing the quantum gravity research group of the Centre de physique théorique in Marseille, France. His books *Seven Brief Lessons on Physics* and *Reality is Not What It Seems* are international bestsellers translated into forty-one languages.

The Order of Time

CARLO ROVELLI

Translated by Erica Segre and Simon Carnell

ALLEN LANE
an imprint of
PENGUIN BOOKS

ALLEN LANE

UK | USA | Canada | Ireland | Australia
India | New Zealand | South Africa

Allen Lane is part of the Penguin Random House group of companies
whose addresses can be found at global.penguinrandomhouse.com

First published in Italy by Adelphi Edizione SPA under the
title *L'ordine del tempo* 2017
This translation first published 2018
004

Set in 11.75/14 pt Garamond MT Std
Typeset by Jouve (UK), Milton Keynes
Printed in Great Britain by Clays Ltd, St Ives plc

A CIP catalogue record for this book is available from the British Library

ISBN: 978–0–241–29252–5

www.greenpenguin.co.uk

MIX
Paper from
responsible sources
FSC® C018179

Penguin Random House is committed to a
sustainable future for our business, our readers
and our planet. This book is made from Forest
Stewardship Council® certified paper.

For Ernesto, Bilo and Edoardo

Contents

Perhaps Time is the Greatest Mystery 1

PART ONE

The Crumbling of Time

1. Loss of Unity 9
2. Loss of Direction 18
3. The End of the Present 34
4. Loss of Independence 52
5. Quanta of Time 72

PART TWO

The World without Time

6. The World is Made of Events, not Things 85
7. The Inadequacy of Grammar 93
8. Dynamics as Relation 102

PART THREE
The Sources of Time

9. Time is Ignorance 115
10. Perspective 125
11. What Emerges from a Particularity 138
12. The Scent of the Madeleine 148
13. The Sources of Time 167

The Sister of Sleep 176

Image Credits 183
Notes 185
Index 207

The verses that open each chapter, unless otherwise indicated, are from versions of Horace's *Odes* translated by Giulio Galetto and published in a charming small volume entitled *In questo breve cerchio* (Edizioni del Paniere, Verona, 1980); English translations by Erica Segre and Simon Carnell.

Perhaps Time is the Greatest Mystery

> Even the words that we are speaking now
> thieving time
> has stolen away,
> and nothing can return. (I, 11)

I stop and do nothing. Nothing happens. I am thinking about nothing. I listen to the passing of time.

This is time, familiar and intimate. We are taken by it. The rush of seconds, hours, years that hurls us towards life then drags us towards nothingness . . . We inhabit time as fish live in water. Our being is being in time. Its solemn music nurtures us, opens the world to us, troubles us, frightens and lulls us. The universe unfolds into the future, dragged by time, and exists according to the order of time.

In Hindu mythology, the river of the cosmos is portrayed with the sacred image of Shiva dancing: his dance supports the coursing of the universe; it is itself the flowing of time. What could be more universal and obvious than this *flowing*?

And yet things are somewhat more complicated than this. Reality is often very different than it seems. The Earth appears to be flat but is in fact spherical.

The sun seems to revolve in the sky when it is really we who are spinning. Neither is the structure of time what it seems to be: it is different from this uniform, universal flowing. I discovered this, to my utter astonishment, in the physics books I read as a university student: time works quite differently from the way it seems to.

In those same books I also discovered that we still don't know how time actually works. The nature of time is perhaps the greatest remaining mystery. Curious threads connect it to those other great open mysteries: the nature of mind, the origin of the universe, the fate of black holes, the very functioning of life on Earth. Something essential continues to draw us back to the nature of time.

Wonder is the source of our desire for knowledge,[1] and the discovery that time is not what we thought it was opens up a thousand questions. The nature of time has been at the centre of my life's work in theoretical physics. In the following pages I give an account of what we have understood about time and the paths that are being followed in our search to understand it better. As well as an account of what we have yet to understand and what it seems to me that we are just beginning to glimpse.

Why do we remember the past and not the future? Do we exist in time, or does time exist in us? What does it really mean to say that time 'passes'? What ties time to our nature as persons, to our subjectivity?

What am I listening to when I listen to the passing of time?

This book is divided into three unequal parts. In the first, I summarize what modern physics has understood about time. It is like holding a snowflake in your hands: gradually, as you study it, it melts between your fingers and vanishes. We conventionally think of time as something simple and fundamental that flows uniformly, independently from everything else, from the past to the future, measured by clocks and watches. In the course of time, the events of the universe succeed each other in an orderly way: pasts, presents, futures. The past is fixed, the future open . . . And yet all of this has turned out to be false.

One after another, the characteristic features of time have proved to be approximations, mistakes determined by our perspective, just like the flatness of the Earth or the revolving of the sun. The growth of our knowledge has led to a slow disintegration of our notion of time. What we call 'time' is a complex collection of structures,[2] of layers. Under increasing scrutiny, in ever greater depth, time has lost layers one after another, piece by piece. The first part of this book gives an account of this crumbling of time.

The second part describes what we have been left with: an empty, windswept landscape almost devoid of all trace of temporality. A strange, alien world which is nevertheless still the one to which we belong. It is like arriving in the high mountains, where there is nothing but snow, rocks and sky. Or like it must have been for Armstrong and Aldrin when venturing on to the motionless sand of the moon. A world stripped to its essence, glittering with an arid and troubling beauty. The physics on which I work – quantum gravity – is an attempt to understand and lend coherent meaning to this extreme and beautiful landscape. To the world without time.

The third part of the book is the most difficult, but also the most vital and the one that most closely involves us. In a world without time, there must still be something that gives rise to the time that we are accustomed to, with its order, with its past that is different from the future, with its smooth flowing. Somehow, our time must emerge around us, at least *for* us and at our scale.[3]

This is the return journey, back towards the time lost in the first part of the book when pursuing the elementary grammar of the world. As in a crime novel, we are now going in search of a guilty party: the culprit who has created time. One by one, we discover the constituent parts of the time that is familiar to us, not now as elementary structures of reality but rather as useful approximations used by the clumsy and bungling mortal creatures we are: aspects of our perspective, and aspects, too, perhaps, that are decisive in determining what we are. Because the mystery of time is ultimately, perhaps, more about ourselves than about the cosmos. Perhaps, as in the first and greatest of all detective novels, Sophocles' *Oedipus Rex*, the culprit turns out to be the detective.

Here, the book becomes a fiery magma of ideas, sometimes illuminating, sometimes confusing. If you decide to follow me, I will take you to where I believe our knowledge of time has reached: up to the brink of that vast nocturnal and star-studded ocean of all that we still don't know.

PART ONE
The Crumbling of Time

1. Loss of Unity

Dances of love intertwine
such graceful girls
lit by the moon
on these clear nights. (I, 4)

The Slowing Down of Time

Let's begin with a simple fact: time passes faster in the mountains than it does at sea level.

The difference is small, but it can be measured with precision timepieces that can be bought today on the internet for a few thousand pounds. With practice, anyone can witness the slowing down of time. With the timepieces of specialized laboratories, this slowing down of time can be detected between levels just a few centimetres apart: a clock placed on the floor runs a little more slowly than one on a table.

It is not just the clocks that slow down: lower down, all processes are slower. Two friends separate, with one of them living in the plains and the other going to live in the mountains. They meet up again years later: the one who has stayed down has lived less, aged less, the mechanism of his cuckoo clock has oscillated fewer times. He has had less time to do things, his plants have grown less, his thoughts have had less time to unfold . . . Lower down, there is simply less time than at altitude.

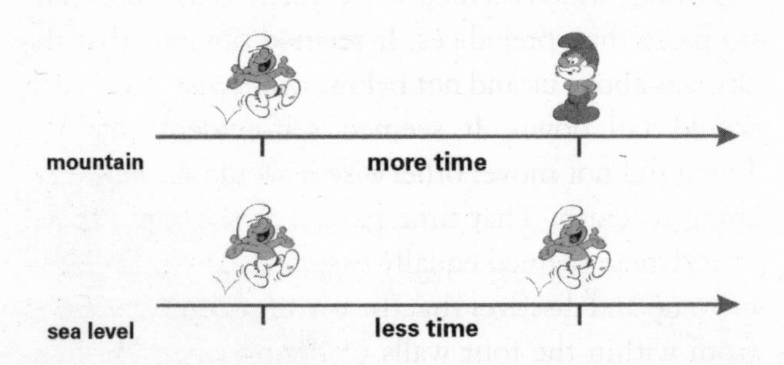

Is this surprising? Perhaps it is. But this is how the world works. Time passes more slowly in some places, more rapidly in others.

The surprising thing, perhaps, is that someone understood this slowing down of time a century before we had clocks precise enough to measure it. His name, of course, was Albert Einstein.

The ability to understand something before it's observed is at the heart of scientific thinking. In antiquity, Anaximander understood that the sky continues beneath our feet long before ships had circumnavigated the Earth. At the beginning of the modern era, Copernicus understood that the Earth turns long before astronauts had seen it do so from the moon. In a similar way, Einstein understood that time does not pass uniformly everywhere *before* the development of clocks accurate enough to measure the different speeds at which it passes.

In the course of making such strides, we learn that the things which seemed self-evident to us were really no more than prejudices. It seemed obvious that the sky was above us and not below; otherwise, the Earth would fall down. It seemed self-evident that the Earth did not move; otherwise it would cause everything to crash. That time passed at the same speed everywhere seemed equally *obvious* to us . . . Children grow up and discover that the world is not as it seemed from within the four walls of their homes. Humankind as a whole does the same.

Einstein asked himself a question which has perhaps puzzled many of us when studying the force of gravity: how can the sun and the Earth 'attract' each other without touching and without utilizing anything between them?

He looked for a plausible explanation and found one by imagining that the sun and the Earth do not attract each other directly but that each of the two gradually acts on that which is between them. And, since what lies between them is only space and time, he imagined that the sun and the Earth each modified the space and time that surrounded them, just as a body immersed in water displaces the water around it. This modification of the structure of time influences in turn the movement of bodies, causing them to 'fall' towards each other.[1]

What does it mean, this 'modification of the structure of time'? It means precisely the slowing down of time described above: a mass slows down time around itself. The Earth is a large mass and slows down time in its vicinity. It does so more in the plains and less in the mountains, because the plains are closer to it. This is why the friend who stays at sea level ages more slowly.

If things fall, it is due to this slowing down of time. Where time passes uniformly, in interplanetary space, things do not fall. They float, without falling. Here on the surface of our planet, on the other hand, the movement of things inclines naturally towards where time passes more slowly, as when we run down the beach into the sea and the resistance of the water on our legs makes us fall headfirst into the waves. Things fall downwards because, down there, time is slowed by the Earth.[2]

Hence, even though we cannot easily observe it, the slowing down of time nevertheless has crucial effects: things fall because of it, and it allows us to keep our feet firmly on the ground. If our feet adhere to the pavement, it is because our whole body inclines naturally to where time runs more slowly – and time passes more slowly for your feet than it does for your head.

Does this seem strange? It is like when, while

watching the sun going down gloriously at sunset, disappearing slowly behind distant clouds, we suddenly remember that it's not the sun that's moving but the Earth that's spinning, and we see with the unhinged eye of the mind our entire planet – and ourselves with it – rotating backwards, away from the sun. We are seeing with 'mad' eyes, like those of Paul McCartney's Fool on the Hill: the crazed vision that sometimes sees further than our bleary, customary eyesight.

Ten Thousand Dancing Shivas

I have an enduring passion for Anaximander, the Greek philosopher who lived twenty-six centuries ago and understood that the Earth floats in space, supported by nothing.[3] We know of Anaximander's thought from other writers. Only one small original fragment of his writings has survived – just one:

> Things are transformed one into another according to necessity, and render justice to one another according to the order of time.

'According to the order of time' (κατὰ τὴν τοῦ κρόνου τάξιν). From one of the crucial, initial moments of natural science there remains nothing but these obscure, arcanely resonant words, this appeal to the 'order of time'.

Astronomy and physics have developed since by following this seminal lead given by Anaximander: by understanding how phenomena occur *according to the order of time.* In antiquity, astronomy described the movements of stars *in time.* The equations of physics describe how things change *in time.* From the equations of Newton, which establish the foundations of mechanics, to those of Maxwell for electromagnetic phenomena; from Schrödinger's equation describing how quantum phenomena evolve, to those of quantum field theory for the dynamics of subatomic particles: the whole of our physics, and science in general, is about how things develop 'according to the order of time'.

It has long been the convention to indicate this time in equations with the letter *t* (the word for time begins with *t* in Italian, French and Spanish, but not in German, Arabic, Russian or Mandarin). What does this *t* stand for? It stands for the number measured by a clock. The equations tell us how things change as the time measured by a clock passes.

But if different clocks mark different times, as we have seen above, what does *t* indicate? When the two friends meet up again after one has lived in the mountains and the other at sea level, the watches on their wrists will show different times. Which of the two is *t*? In a physics laboratory, a clock on a table and another on the ground run at different speeds.

Which of the two tells the time? How do we describe the difference between them? Should we say that the clock on the ground has slowed relative to the real time recorded on the table? Or that the clock on the table runs faster than the real time measured on the ground?

The question is meaningless. We might just as well ask what is *most* real – the value of sterling in dollars or the value of dollars in sterling. There is no 'truer' value; they are two currencies which have value *relative to each other*. There is no truer time. There are two times that change *relative to each other*. Neither is truer than the other.

But there are not just *two* times. Times are legion: a different one for every point in space. There is not one single time; there is a vast multitude of them.

The time indicated by a particular clock measuring a particular phenomenon is called 'proper time' in physics. Every clock has its proper time. Every phenomenon that occurs has its proper time, its own rhythm.

Einstein has given us the equations that describe how proper times develop *relative to each other*. He has shown us how to calculate the difference between two times.[4]

The single quantity 'time' melts into a spiderweb of times. We do not describe how the world evolves in time: we describe how things evolve in local time, and how local times evolve *relative to each other.* The world is not like a platoon advancing at the pace of a single commander. It's a network of events affecting each other.

This is how time is depicted in Einstein's general theory of relativity. His equations do not have a single 'time'; they have innumerable times. Between two events, just as between the two clocks that are separated and then brought together again, the duration is not a single one.[5] Physics does not describe how things evolve 'in time' but how things evolve in their own times, and how 'times' evolve relative to each other.*

Time has lost its first aspect or layer: its unity. It has

* Grammatical note. The word 'time' has several meanings linked to each other but distinct from one another: 1. 'Time' is the general phenomenon of the succession of events ('The inaudible and noiseless foot of time'); 2. 'Time' indicates an interval in this succession ('To morrow, and to morrow, and to morrow, / Creeps in this petty pace from day to day, / To the last syllable of recorded time'); or 3. its duration ('O gentlemen, the time of life is short'); 4. 'Time' can also indicate a particular moment ('Time will come and take my love away'), often the current one ('The time is out of joint'); 5. 'Time' indicates the variable that measures duration ('Acceleration is the derivative of speed with respect to time'). In this book, I employ each of these meanings freely, just as in common usage. In case of any confusion, please refer back to this note.

a different rhythm in every different place and passes here differently from there. The things of this world interweave dances made to different rhythms. If the world is upheld by the dancing Shiva, there must be ten thousand such dancing Shivas, like the dancing figures painted by Matisse . . .

2. Loss of Direction

If more gently than Orpheus
who moved even the trees
you were to pluck the zither
the life-blood would not return
to the vain shadow . . .
Harsh fate,
but its burden becomes lighter
to bear, since everything
that attempts to turn back
is impossible. (I, 24)

Where Does the Eternal Current Come From?

Clocks may well run at different speeds in the mountains and in the plains, but is this really what concerns us, ultimately, about time? In a river, the water flows more slowly near to its banks, faster in the middle – but it is still flowing . . . Is time not also something that always flows – from the past to the future? Let's leave aside the precise measurement of *how much* time passes that we wrestled with in the preceding chapter: the *numbers* by which time is measured. There's another, more essential aspect to time: its passage, its flow, the *eternal current* of the first of Rilke's *Duino Elegies*:

> The eternal current
> Draws all the ages along with it
> Through both realms,
> Overwhelming them in both.[1]

Past and future are different from each other. Cause precedes effect. Pain comes after a wound, not before it. The glass shatters into a thousand pieces, and the pieces do not re-form into a glass. We cannot change the past; we can have regrets, remorse, memories. The future instead is uncertainty, desire, anxiety, open space, destiny, perhaps. We can live towards it, shape it, because it does not yet exist. Everything is still possible . . . Time is not a line with two equal directions: it is an arrow with different extremities.

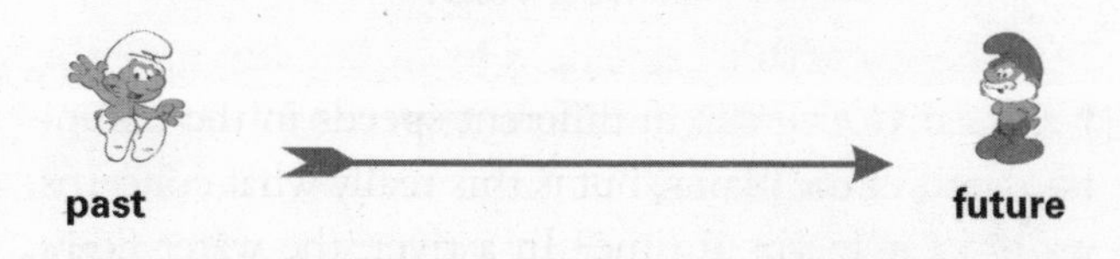

And it is this, rather than the speed of its passing, that matters most to us about time. This is the fundamental thing about time. The secret of time lies in this slippage that we feel on our pulse, viscerally, in the enigma of memory, in anxiety about the future. This is what it means to think about time. What exactly is this *flowing*? Where is it nestled in the grammar of the world? What distinguishes the past, its having been, from the future, its not having been yet, in the folds of the mechanism of the world? Why, to us, is the past so different from the future?

Nineteenth- and twentieth-century physics engaged with these questions and ran into something unexpected and disconcerting – much more so than the relatively marginal fact that time passes at different speeds in different places. The difference between past and future, between cause and effect, between memory and hope, between regret and intention . . . in the elementary laws that describe the mechanisms of the world, there is no such difference.

Heat

It all began with a regicide. On 16 January 1793 the National Convention in Paris sentenced Louis XVI to death. Rebellion is perhaps among the deepest roots of science: the refusal to accept the present order of things.[2] Among those who took the fatal decision was a friend of Robespierre called Lazare Carnot. Carnot had a passion for the great Persian poet Saadi Shirazi. Captured and enslaved at Acre by the Crusaders, Shirazi is the author of those luminous verses that now stand at the entrance of the headquarters of the United Nations:

> All of the sons of Adam are part of
> one single body,
> They are of the same essence.
> When time afflicts us with pain
> In one part of that body
> All the other parts feel it too.
> If you fail to feel the pain of others
> You do not deserve the name of man.

Perhaps poetry is another of science's deepest roots: the capacity to see beyond the visible. Carnot names his first son after Saadi. Sadi Carnot is thus born out of poetry and rebellion.

As a young man, he develops a passion for those steam engines that at the start of the nineteenth century are beginning to transform the world by using fire to make things turn. In 1824 he writes a pamphlet with the alluring title 'Reflections on the Motive Power of Fire' in which he seeks to understand the theoretical basis of the functioning of these machines. The little treatise is packed with mistaken assumptions: he imagines that heat is a concrete entity – a kind of fluid that produces energy by 'falling' from hot things to cold, just as the water in a waterfall produces energy by falling from above to below. But it contains a key idea: that steam engines function, in the final analysis, because the heat passes from hot to cold.

Sadi's pamphlet finds its way into the hands of a fierce-eyed, austere Prussian professor called Rudolf Clausius. It is he who grasps the fundamental issue at stake, formulating a law that was destined to become famous: if nothing else around it changes, heat *cannot* pass from a cold body to a hot one.

The crucial point here is the difference from what happens with falling bodies: a ball may fall, but it can also come back up, by rebounding, for instance. Heat cannot.

This is the *only* basic law of physics that distinguishes the past from the future.

None of the others do so. Not Newton's laws governing the mechanics of the world; not the equations for electricity and magnetism formulated by Maxwell. Not Einstein's on relativistic gravity, nor those of quantum mechanics devised by Heisenberg, Schrödinger and Dirac. Not those for elementary particles formulated by twentieth-century physicists . . . *Not one* of these equations distinguishes the past from the future.[3] If a sequence of events is allowed by these equations, so is the same sequence run backwards in time.[4] In the elementary equations of the world,[5] the arrow of time appears *only* where there is heat.* The link between time and heat is therefore fundamental: every time a difference is manifested between the past and the future, heat is involved. In every sequence of events that becomes absurd if projected backwards, there is something that is heating up.

If I watch a film that shows a ball rolling, I cannot tell if the film is being projected correctly or in reverse. But, if the ball stops, I know that it is being run properly; run backwards, it would show an implausible event: a ball starting to move by itself. The ball's slowing down and coming to rest are due to friction, and friction produces heat. Only where there is heat is there a distinction between past and future. Thoughts, for instance, unfold from the past to the future, not vice versa – and, in fact, thinking produces heat in our heads . . .

Clausius introduces a quantity that measures this

* Strictly speaking, the arrow of time can also manifest itself in phenomena that are not linked directly to heat but share crucial aspects with it, for instance in the use of retarded potentials in electrodynamics. What follows applies also for these phenomena – in particular, the conclusions. I prefer here not to overload the discussion by breaking it down into all its different sub-cases.

irreversible progress of heat in only one direction and, since he was a cultivated German, he gives it a name taken from ancient Greek, *entropy*:

> I prefer to take the names of important scientific quantities from ancient languages, so that they may be the same in all the living languages. I therefore propose to call entropy the quantity (*S*) of a body, from the Greek word for transformation: ἡτροπή.[6]

Clausius's entropy, indicated by the letter S, is a measurable and calculable[7] quantity that increases or remains the same but *never decreases*, in an isolated process. In order to indicate that it never decreases, we write:

$$\Delta S \geq 0$$

This reads: 'Delta S is always greater than or equal to zero', and we call this 'the second principle of thermodynamics' (the first being the conservation of energy). Its nub is the fact that heat passes only from hot bodies to cold, never the other way round.

Forgive me for the equation – it's the only one in the book. It is the equation for time's arrow, and I could hardly refrain from including it in my book about time.

390

so erhält man die Gleichung:

$$(64)\quad \int \frac{dQ}{T} = S - S_0,$$

welche, nur etwas anders geordnet, dieselbe ist, wie die unter (60) angeführte zur Bestimmung von S dienende Gleichung.

Sucht man für S einen bezeichnenden Namen, so könnte man, ähnlich wie von der Gröfse U gesagt ist, sie sey der *Wärme- und Werkinhalt* des Körpers, von der Gröfse S sagen, sie sey der *Verwandlungsinhalt* des Körpers. Da ich es aber für besser halte, die Namen derartiger für die Wissenschaft wichtiger Gröfsen aus den alten Sprachen zu entnehmen, damit sie unverändert in allen neuen Sprachen angewandt werden können, so schlage ich vor, die Gröfse S nach dem griechischen Worte *ἡ τροπή*, die Verwandlung, die *Entropie* des Körpers zu nennen. Das Wort *Entropie* habe ich absichtlich dem Worte *Energie* möglichst ähnlich gebildet, denn die beiden Gröfsen, welche durch diese Worte benannt werden sollen, sind ihren physikalischen Bedeutungen nach einander so nahe verwandt, dafs eine gewisse Gleichartigkeit in der Benennung mir zweckmäfsig zu seyn scheint.

Fassen wir, bevor wir weiter gehen, der Uebersichtlichkeit wegen noch einmal die verschiedenen im Verlaufe der Abhandlung besprochenen Gröfsen zusammen, welche durch die mechanische Wärmetheorie entweder neu eingeführt sind, oder doch eine veränderte Bedeutung erhalten haben, und welche sich alle darin gleich verhalten, dafs sie durch den augenblicklich stattfindenden Zustand des Körpers bestimmt sind, ohne dafs man die Art, wie der Körper in denselben gelangt ist, zu kennen braucht, so sind es folgende sechs: 1) der *Wärmeinhalt*, 2) der *Werkinhalt*, 3) die Summe der beiden vorigen, also der *Wärme- und Werkinhalt* oder die *Energie*; 4) der *Verwandlungswerth des Wärmeinhaltes*, 5) die *Disgregation*, welche als der Verwandlungswerth der stattfindenden Anordnung der Bestandtheile zu

This is the page of the article by Clausius in which he introduces for the first time the concept and the word 'entropy'. The equation provides the mathematical definition of the variation of entropy (S–S_0) of a body: the sum (integral) of the quantity of heat dQ leaving the body at the temperature T.

It is the only equation of fundamental physics that knows any difference between past and future. The only one that speaks of the flowing of time. Behind this unusual equation, an entire world lies hidden.

Revealing it will fall to an unfortunate and engaging Austrian, the grandson of a watchmaker, a tragic and romantic figure, Ludwig Boltzmann.

Blur

It is Boltzmann who begins to see what lies behind the equation $\Delta S \geq 0$, throwing us into one of our most dizzying dives towards understanding the intimate grammar of our world.

Boltzmann works in Graz, Heidelberg, Berlin, Vienna, and then in Graz again. He liked to attribute his restlessness to the fact that he was born during Mardi Gras. He was only partly joking, since the instability of his character was real enough, oscillating as it did between elation and depression. He was short and stout, with dark, curly hair and the beard of a Taliban; his girlfriend called him 'my dear sweet chubby one'. It was he, this Ludwig, who was the luckless hero of time's directionality.

Sadi Carnot thought that heat was a substance, a fluid. He was wrong. Heat is the microscopic agitation of molecules. Hot tea is tea in which the molecules are very agitated. Cold tea is tea in which the molecules are only a little agitated. In an ice cube, warming up and melting molecules become increasingly agitated and lose their strict connections.

At the end of the nineteenth century there were many who still did not believe in the existence of molecules and atoms: Ludwig was convinced of their reality and entered the fray on behalf of his belief. His diatribes against those who doubted the reality of atoms became legendary. 'Our generation were at heart all on his side,' remarked one of the young lions of quantum mechanics, years later.[8] In one of these fiery polemics, at a conference in Vienna, a noted physicist[9] maintained against him that scientific materialism was dead because the laws of matter are not subject to the directionality of time. Physicists are not immune from talking nonsense.

Looking at the sun going down, the eyes of Copernicus had seen the world turning. Looking at a glass of still water, the eyes of Boltzmann saw atoms and molecules frenziedly *moving*.

We see the water in a glass like the astronauts saw the Earth from the moon: calm, gleaming, blue. From the moon, they could see nothing of the exuberant agitation of life on Earth, its plants and animals, desires and despairs. Only a veined blue ball. Within the reflections in a glass of water, there is an analogous tumultuous life, made up of the activities of a myriad of molecules – many more than there are living beings on Earth.

This tumult *stirs up* everything. If one section of the molecules is still, it becomes stirred up by the frenzy of neighbouring ones that set them in motion, too: the agitation spreads, the molecules bump into and shove each other. In this way, cold things are heated in contact with hot ones: their molecules become jostled by hot ones and pushed into ferment. That is, they heat up.

Thermal agitation is like a continual shuffling of a pack of cards: if the cards are in order, the shuffling disorders them. In this way, heat passes from hot to cold, and not vice versa: by shuffling, by the natural disordering of everything. The growth of entropy is nothing other than the ubiquitous and familiar natural increase of disorder.

This is what Boltzmann understood. The difference between past and future does not lie in the elementary laws of motion; it does not reside in the deep grammar of nature. It is the natural disordering that leads to gradually less particular, less special situations.

It was a brilliant intuition, and a correct one. But does it clarify the difference between past and future? It does not. It just shifts the question. The question now becomes: why, in one of the two directions of time – the one we call past – were things more ordered? Why was the great pack of cards of the universe in order in the past? Why, in the past, was entropy lower?

If we observe a phenomenon that *begins* in a state of lower entropy, it is clear why entropy increases – because in the process of reshuffling everything becomes disordered. But why do the phenomena that we observe around us in the cosmos *begin* in a state of lower entropy in the first place?

Here we get to the key point. If the first twenty-six cards in a pack are all red and the next twenty-six are all black, we say that the configuration of the cards is 'particular'; that it is 'ordered'. This order is lost when the pack is shuffled. The initial ordered configuration is a configuration 'of low entropy'. But notice that it is particular if we look at the *colour* of the cards – red or black. It is particular because I am looking at the colour. Another configuration will be particular if the first twenty-six cards consist of only hearts and spades. Or if they are all odd numbers, or the twenty-six most creased cards in the pack, or exactly the same twenty-six of three days ago . . . Or if they share any other characteristic. If we think about it carefully, *every configuration is particular*, every configuration is singular, if we look at *all* of its details, since every configuration always has something about it that characterizes it in a unique way. Just as, for its mother, every child is particular and unique.

It follows that the notion of certain configurations being more particular than others (twenty-six red cards followed by twenty-six black, for example) makes sense only if I limit myself to noticing only certain aspects of the cards (in this case, the colours). If I distinguish between all the cards, the configurations are all equivalent: none of them is more or less particular than others.[10] The notion of 'particularity' is born only at the moment we begin to see the universe in a blurred and approximate way.

Boltzmann has shown that entropy exists because we describe the world in a blurred fashion. He has demonstrated that entropy is precisely the quantity that counts *how many* are the different configurations that our blurred vision does *not* distinguish between. Heat, entropy and the lower entropy of the past are notions that belong to an approximate, statistical description of nature.

The difference between past and future is deeply linked to this blurring . . . So, if I could take into account all the details of the exact, microscopic state of the world, would the characteristic aspects of the flowing of time disappear?

Yes. If I observe the microscopic state of things, then the difference between past and future vanishes. The future of the world, for instance, is determined by its present state – though neither more nor less than is the past.[11] We often say that causes precede effects and yet, in the elementary grammar of things, there is no distinction between 'cause' and 'effect'.* There are regularities, represented by what we call physical laws, that link events of different times, but they are symmetric between future and past. In a

* There are a few more details given on this point in Chapter 11.

microscopic description, there can be no sense in which the past is different from the future.*

This is the disconcerting conclusion that emerges from Boltzmann's work: the difference between the past and the future refers only to *our own* blurred vision of the world. It's a conclusion that leaves us flabbergasted: is it really possible that a perception so vivid, basic, existential – my perception of the passage of time – depends on the fact that I cannot apprehend the world in all of its minute detail? On a kind of distortion that's produced by myopia? Is it true that, if I could see exactly and take into consideration the actual dance of millions of molecules, then the future would be 'just like' the past? Is it possible that I have as much knowledge of the past – or ignorance of it – as I do of the future? Even allowing for the fact that our perceptions of the world are frequently wrong, can the world really be *so* profoundly different from our perception of it as this?

* The point is not that what happens to a cold teaspoon in a cup of hot tea depends on whether I have a blurred vision of it or not. What happens to the spoon and to its molecules as well, obviously, does not depend on how I view it. It just happens, regardless. The point is that the description in terms of heat, temperature and the passage of heat from tea to spoon is a blurred vision of what happens, and that it is only in this blurred vision that a startling difference between past and future appears.

All this undermines the very basis of our usual way of understanding time. It provokes incredulity, just as much as the discovery of the movement of the Earth did. But, just as with the movement of the Earth, the evidence is overwhelming: all the phenomena that characterize the flowing of time are reduced to a 'particular' state in the world's past, the 'particularity' of which may be attributed to the blurring of our perspective.

Later on, I will delve into the mystery of this blurring, to see how it is tied to the strange initial improbability of the universe. For now, I will end with the mind-boggling fact that entropy, as Boltzmann fully understood, is nothing other than the number of microscopic states that our blurred vision of the world fails to distinguish.

The equation which states precisely this[12] is carved on Boltzmann's tomb in Vienna, above a marble bust which portrays him as an austere and surly figure, such as I don't believe he ever was in life. Many young students of physics go to visit his tomb, and linger there to ponder. And sometimes the odd elderly professor of physics as well.

Time has lost another of its crucial components: the intrinsic difference between past and future. Boltzmann understood that there is nothing intrinsic about the flowing of time. That it is only the blurred reflection of a mysterious improbability of the universe at a point in the past.

The source of Rilke's *eternal current* is nothing other than this.

Appointed a university professor at just twenty-five years old; received at court by the Emperor at the apex of his success; severely criticized by the majority of the academic world, which did not understand his ideas; always precariously balanced between enthusiasm and depression: the 'dear sweet chubby one', Ludwig Boltzmann, will end his life by hanging himself.

He does so at Duino, near Trieste, while his wife and daughter are swimming in the Adriatic.

The same Duino where, just a few years later, Rilke will write his *Elegies*.

3. The End of the Present

It opens
to this gentle breeze
of Spring
the sealed-in cold
of the still season,
and the boats return to the sea . . .
Now we must braid
crowns with which
to adorn our heads. (I, 4)

Speed Also Slows Down Time

Ten years before understanding that time is slowed down by mass,[1] Einstein had realized that it was slowed down by speed.[2] The consequence of this discovery for our basic intuitive perception of time is the most devastating of all.

The fact itself is quite simple. Instead of sending the two friends from the first chapter to the mountains and the plains, respectively, let's ask one of them to stay still and the other one to walk around. Time passes more slowly for the one who keeps moving.

As before, the two friends experience different durations: the one who moves ages less quickly, his watch marks less time passing; he has less time in which to think; the plant he is carrying takes longer to germinate, and so on. For everything that moves, time passes more slowly.

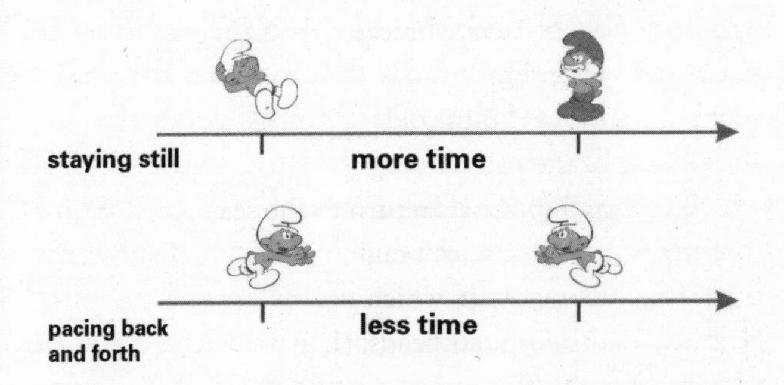

For this effect to become perceptible, one must move very quickly. It was first measured in the 1970s, using precision watches on aeroplanes.[3] The watch on board a plane displays a time behind that displayed by the one on the ground. Today, the slowing down of time can be observed in many physics experiments.

In this case, too, Einstein had understood that time slows down *before* the phenomenon was actually measured – when he was just twenty-five years old and studying electromagnetism.

It turned out to be a not particularly complex deduction. Electricity and magnetism are well described by

Maxwell's equations. These contain the usual time variable t but have a curious property: if you travel at a certain velocity, then *for you* the equations of Maxwell are no longer true (that is, they don't describe what you measure) *unless* you call 'time' a *different* variable: t'.[4] Mathematicians had become aware of this curious feature of Maxwell's equations,[5] but no one had been able to understand what it meant. Einstein grasped its significance: t is the time that passes if I stay still, the rhythm at which things occur that are stationary, like myself; t' is 'your time', the rhythm at which things happen that move with you. t is the time that my watch measures when it is stationary; t' is the time that your watch measures when it is moving. Nobody had imagined previously that time could be different for a stationary watch and one that was being moved. Einstein had read this within the equations of electromagnetism, by taking them seriously.[6]

A moving object therefore experiences a shorter duration than a stationary one: a watch marks fewer seconds, a plant grows more slowly, a young man dreams less. For a moving object,* time contracts.

* 'In motion' in relation to what? How can we determine which of the two objects moves, if the motion is only relative? This is an issue that has confused many. The correct answer (rarely given) is this: in motion relative to the *only* reference in which the point in space where the two clocks separate is the same point in space where they get back together. There is

Not only is there no single time for different places – there is not even a single time for any particular place. A duration can be associated only with the movement of something, with a given trajectory.

'Proper time' depends not only on where you are and your degree of proximity to masses; it depends also on the speed at which you move.

It's a strange enough fact in itself, but its consequences are extraordinary. Hold on tight, because we are about to take off.

'Now' Means Nothing

What is happening 'now' in a distant place? Imagine, for example, that your sister has gone to *Proxima b*, the recently discovered planet that orbits a star at approximately four light years' distance from us. What is your sister doing *now* on *Proxima b*?

The only correct answer is that the question makes no sense. It is like asking 'What is *here*, in Peking?'

---

only a single straight line between two events in spacetime, from A to B: it's the one along which time is maximum, and the speed *relative to this line* is the one that slows time. If the clocks separate and are not brought together again, there is no point asking which one is fast and which one is slow. If they come together, they can be compared, and the speed of each one becomes a well-defined notion.

when we are in Venice. It makes no sense because if I use the word 'here' in Venice, I am referring to a place in Venice, not in Peking.

If your sister is in the room and you ask what she is doing *now*, the answer is usually an easy one: you look at her and you can tell. If she's far away, you phone her and ask what she's doing. But take care: if you look at your sister, you are receiving light that travels from her to your eyes. The light takes time to reach you, let's say a few nanoseconds – a tiny fraction of a second – therefore, you are not quite seeing what she is doing *now* but what she was doing a few nano-seconds ago. If she is in New York and you phone her from Liverpool, her voice takes a few milliseconds to reach you, so the most you can claim to know is what your sister was up to a few milliseconds ago. Not a significant difference, perhaps.

If your sister is on *Proxima b*, however, light takes four years to reach you from there. Hence, if you look at her through a telescope, or receive a radio communication from her, you know what she was doing four years ago rather than what she is doing now. 'Now' on *Proxima b* is definitely not what you see through the telescope, or what you can hear from her voice over the radio.

So perhaps you can say that what your sister is doing *now* is what she will be doing four years after the moment that you see her through the telescope? But no, this does not work: four years after you have seen her through the telescope, in her time, she might already have returned to Earth and could be (yes! this is really possible!) ten terrestrial years in the future. But 'now' cannot be in the future . . .

Perhaps we can do this: if, ten years ago, your sister had left for *Proxima b*, taking with her a calendar to keep track of the passage of time, can we think that *now* for her is when she has recorded that ten years have passed? No, this does not work either: she might have returned here ten of *her* years after leaving, arriving back where, in the meantime, twenty years have elapsed. So when the hell *is* 'now' on *Proxima b*?

The truth of the matter is that we need to give up asking the question.[7]

There is no special moment on *Proxima b* that corresponds to what constitutes the present here and now.

Dear reader, pause for a moment to let this conclusion sink in. In my opinion, it is the most astounding conclusion arrived at in the whole of contemporary physics.

It simply makes no sense to ask which moment in the life of your sister on *Proxima b* corresponds to *now*. It is like asking which football team has won a basketball championship, how much money a swallow has earned, or how much a musical note weighs. They are nonsensical questions because football teams play football, not basketball; swallows do not busy themselves earning money; sounds cannot be weighed. 'Basketball champions' refers to a team of basketball players, not to footballers. Monetary profit refers to human society, not to swallows. The notion of 'the present' refers to things that are close to us, not to anything that is far away.

Our 'present' does not extend throughout the universe. It is like a bubble around us.

How far does this bubble extend? It depends on the precision with which we determine time. If by nanoseconds, the present is defined only over a few metres; if by milliseconds, it is defined over thousands of kilometres. As humans, we distinguish tenths of a second only with great difficulty; we can easily consider our entire planet to be like a single bubble where we can speak of the present as if it were an instant shared by us all. This is as far as we can go.

There is our past: all the events that happened before what we can witness now. There is our future: the events that will happen after the moment from which we can see the here and now. Between this past and this future there is an interval that is neither past nor future and still has a duration: fifteen minutes on Mars; eight years on *Proxima b*; millions of years in the Andromeda galaxy. It is the expanded present.[8] It is perhaps the greatest and strangest of Einstein's discoveries.

The idea that a well-defined *now* exists throughout the universe is an illusion, an illegitimate extrapolation of our own experience.[9]

It is like the point where the rainbow touches the

forest. We think that we can see it – but, if we go to look for it, it isn't there.

If I were to ask, 'Are these two stones at the same height?' in interplanetary space, the correct answer would be: 'It's a question that doesn't make sense, because there isn't a single notion of "same height" throughout the universe.' If I ask whether two events – one on Earth and the other on *Proxima b* – are happening 'at the same moment', the correct answer would be: 'It's a question that doesn't make sense, because there is no such thing as "the same moment" definable in the universe.'

The 'present of the universe' is meaningless.

Temporal Structure without the Present

Gorgo is the woman who saved Greece by realizing that a wax-covered tablet sent there from Persia carried a secret message concealed *beneath* the wax: a message that forewarned the Greeks of a Persian attack. Gorgo had a son called Pleistarchus, fathered by the king of Sparta, the hero of Thermopylae: Leonidas. Leonidas was Gorgo's uncle, the brother of her father, Cleomenes. Who belongs to the 'same generation' as Leonidas? Gorgo, who is the mother of his son – or Cleomenes, who is the son of the same father? Here is a diagram for those who, like me, have difficulties with genealogy:

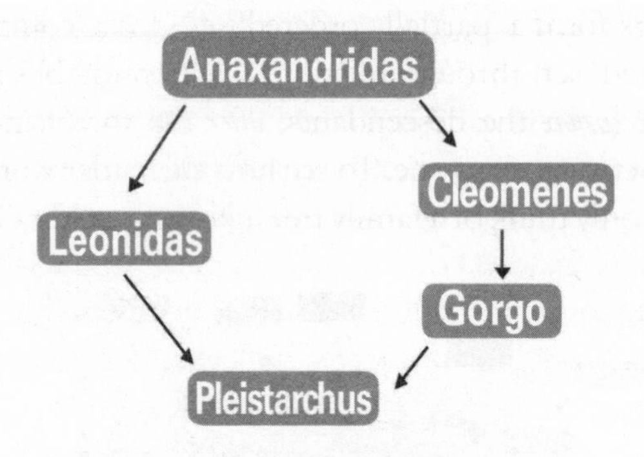

There is an analogy between generations and the temporal structure of the world as revealed by relativity. It makes no sense to ask if it is Cleomenes or Gorgo who is 'of the same generation' as Leonidas, because there is no single concept[10] of 'same generation'. If we say that Leonidas and his brother are 'of the same generation' because they have the same father, and that Leonidas and his wife are 'of the same generation' because they have a son together, we must therefore say that this 'same generation' includes Gorgo and her own father! The filial relationship establishes an order between human beings (Leonidas, Gorgo and Cleomenes come *after* Anaxandridas and *before* Pleistarchus), but not between *any* humans: Leonidas and Gorgo are neither before or after in respect to each other.

Mathematicians have a term for the order established by filiation: 'partial order'. A partial order establishes a relation of *before* and *after* between certain elements, but not between any two of them. Human beings form a 'partially ordered' set (not a 'completely ordered' set) through filiation. Filiation establishes an order (*before* the descendants, *after* the forebears), but not between everyone. To see how this order works we need only think of a family tree, like this one for Gorgo:

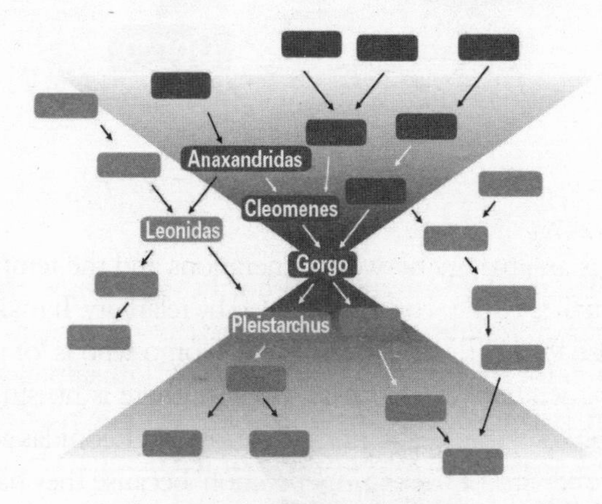

There is a cone-shaped 'past' made up of her forebears, and a 'future' cone comprised of her descendants. Those who are neither ancestors nor descendants remain outside of the cones.

Every human being has their own past cone of ancestors and future cone of descendants. Those of Leonidas are shown below, alongside Gorgo's.

The temporal structure of the universe is very similar to this one. It is also made of cones. The relation of 'temporal precedence' is a partial order made of cones.[11] Special relativity is the discovery that the temporal

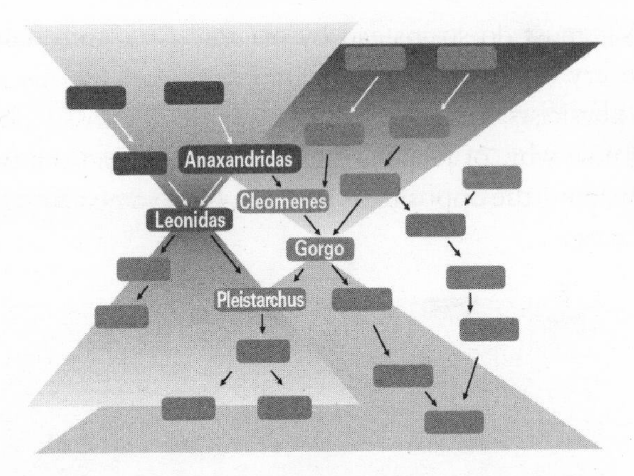

structure of the universe is like the one established by filiation: it defines an order between the events of the universe that is *partial*, not *complete*. The expanded present is the set of events that are neither past nor future: it exists, just as there are human beings who are neither our descendants nor our forebears.

If we want to represent all the events in the universe and their temporal relations, we can no longer do so with a single, universal distinction between past, present and future, like this:

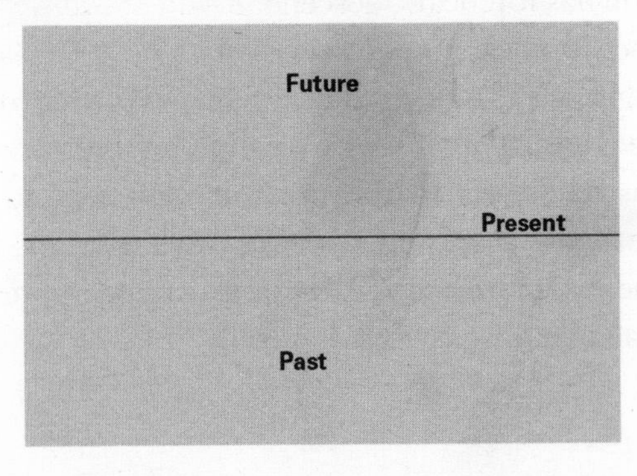

We must do so instead by placing above and below every event the cones of its future and past events (physicists have the habit in such diagrams, I don't know why, of placing the future above and the past below – the opposite of how it is done in genealogical trees):

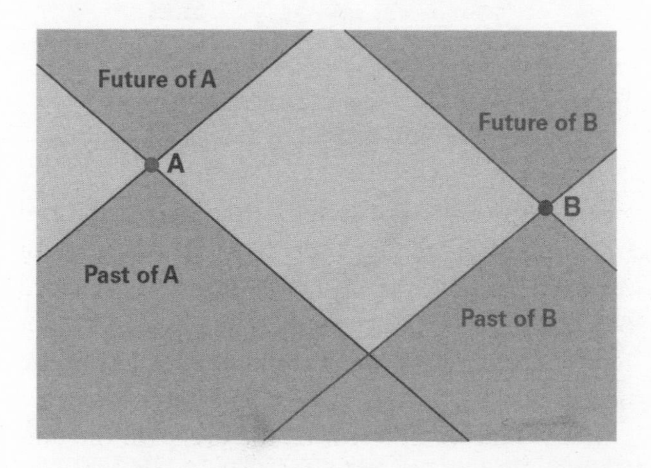

Every event has its past, its future and a part of the universe that is neither past nor future, just as every person has forebears, descendants and others who are neither forebears nor descendants.

Light travels along the oblique lines that delimit these cones. This is why we call them 'light cones'. It is customary, as in the previous diagram, to draw these lines at an angle of forty-five degrees, but it would be more realistic to make them more horizontal, like this:

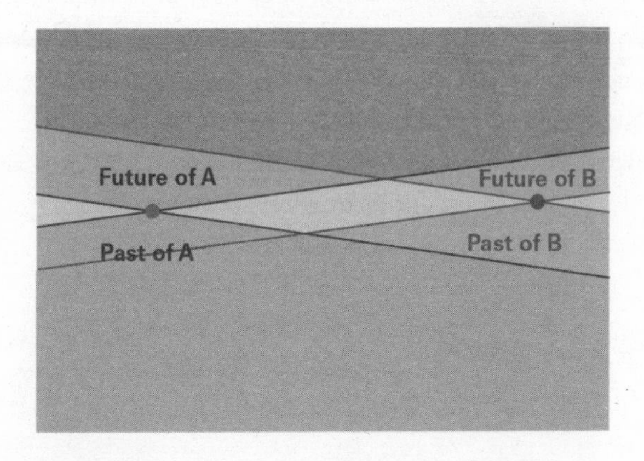

The reason for this is that, at the scale to which we are accustomed, the expanded present separating our past from our future is extremely brief (a matter of nanoseconds) and almost imperceptible, as a result of which it is 'squashed' into a thin horizontal band we usually call 'the present', without any qualification.

In short, a common present does not exist: the temporal structure of spacetime is not a stratification of times such as this:

time n
⋮
time 3
time 2
time 1

It is, rather, a structure made up entirely of light cones:

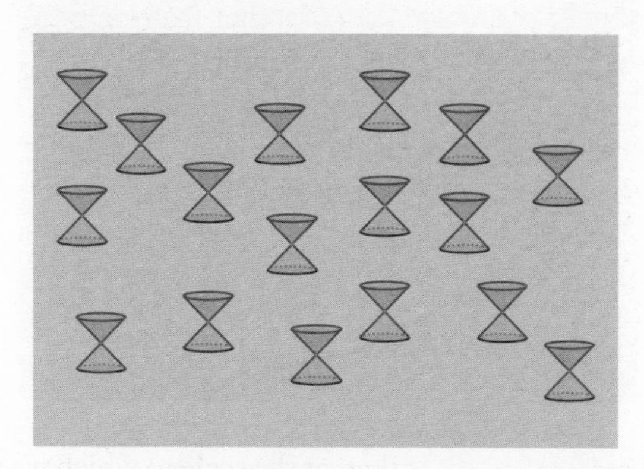

This is the structure of spacetime that Einstein understood when he was twenty-five years old.

Ten years later, he comes to understand that the speed at which time flows changes from place to place. It follows that spacetime does not really have the order outlined above but can be distorted. It now looks rather more like this:

When a gravitational wave passes, for example, the small light cones oscillate together from right to left, like ears of wheat blown by the wind.

The structure of the cones can even be such that, advancing always towards the future, one can return to the same point in spacetime, like this:

In this way, a continuous trajectory towards the future returns to the originating event, to where it began.*[12] The first to realize this was Kurt Gödel, the great

* The 'closed temporal lines', where the future returns us to the past, are the ones that frighten those who imagine that a son could go on to kill his mother before his own birth. But there is no logical contradiction entailed by the existence of closed temporal lines or journeys to the past; we are the ones who complicate things with our confused fantasies about the supposed freedom of the future.

twentieth-century logician who was Einstein's last friend, accompanying him on walks along the streets of Princeton.

Near to a black hole, the lines converge towards it, like this:[13]

This is because the mass of the black hole slows time to such a degree that, at its border (called the 'horizon'), time stands still. If you look closely, you will see that the surface of the black hole is parallel to the edges of the cones. So, in order to exit from a black hole, you would need to move (like the trajectory marked in black in the following diagram) towards the present rather than towards the future!

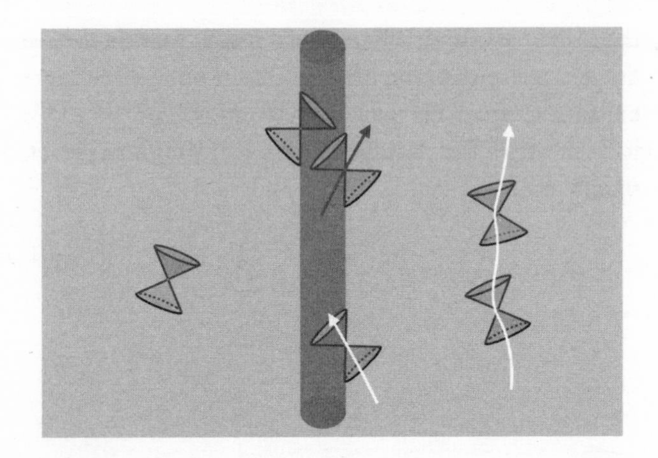

This is impossible. Objects can only move towards the future, as in the trajectories shown in the diagram in white. This is what constitutes a black hole: an inclination of the light cones towards the interior, marking a horizon, closing off a region of space in the future from everything that surrounds it. It is nothing other than this. It is the curious local structure of the present that produces black holes.

More than a hundred years have passed since we learned that the 'present of the universe' does not exist. And yet this continues to confound us and still seems difficult to conceptualize. Every so often a physicist mutinies and tries to show that it isn't true.[14] Philosophers continue to discuss the disappearance of the present. Today, there are often conferences devoted to the subject.

If the present has no meaning, then what 'exists' in

the universe? Is not what 'exists' precisely what is here 'in the present'? The whole idea that the universe exists *now* in a certain configuration and changes together with the passage of time simply doesn't stack up any more.

4. Loss of Independence

And on that wave
we will all have to navigate,
all who are nourished
by the fruits of the Earth. (II, 14)

What Happens when Nothing Happens?

It only takes a few micrograms of LSD to expand our experience of time on to an epic and magical scale.[1] 'How long is forever?' asks Alice. 'Sometimes, just one second,' replies the White Rabbit. There are dreams lasting an instant in which everything seems frozen for an eternity.[2] Time is elastic in our personal experience of it. Hours fly by like minutes, and minutes are oppressively slow, as if they were centuries. On the one hand, time is structured by the liturgical calendar: Easter follows Lent, and Lent is followed by Christmas; Ramadan opens with Hilal, and closes with Eid al-Fitr. On the other, every mystical experience, such as the sacred moment in which the host is consecrated, throws the faithful outside of time, putting them in touch with eternity. Before Einstein told us that it wasn't true, how the devil did we get it into our heads that time passes everywhere at the same speed? It was certainly not our direct *experience* of the passage of time that gave us the idea that time elapses at the same rate, always and everywhere. So where did we get it from?

For centuries, we have *divided* time into days. The word 'time' derives from an Indo-European root – *di* or *dai* – meaning 'to divide'. For centuries, we have divided the days into hours.[3] For most of those centuries, however, hours were longer in the summer and shorter in the winter, because the twelve hours divided the time between dawn and sunset: the first hour was dawn, and the twelfth was sunset, regardless of the season, as we read in the parable of the winegrower in the Gospel according to Matthew.[4] Since, as we say nowadays, during summer 'more time' passes between dawn and sunset than during the winter, in the summer the hours were longer, and the hours were shorter in wintertime . . .

Sundials, hourglasses and water clocks already existed in the ancient world, in the Mediterranean region and in China – but they did not play the cruel role that clocks have today in the organization of our lives. It is only in the fourteenth century in Europe that people's lives start to be regulated by mechanical clocks. Cities and villages build their churches, erect bell-towers next to them and place a clock on the bell-tower to mark the rhythm of collective activities. The era of clock-regulated time begins.

Gradually, time slips from the hands of the angels and into those of the mathematicians – as is graphically illustrated at Strasbourg Cathedral, where two sundials are surmounted, respectively, by an angel (one inspired by earlier sundials from 1200) and by a mathematician (on the sundial put there in 1400).

The usefulness of clocks supposedly resides in the fact that they tell the same time. And yet this idea is also more modern than we might imagine. For centuries, as long as travel was on horseback, on foot or in carriages, there was no reason to synchronize clocks between one place and another. There was good reason for *not* doing so. Midday is, by definition, when the sun is at its highest. Every city and village had a sundial that registered the moment the sun was at its midpoint, allowing the clock on the bell-tower to be regulated with it, for all to see. But the sun does not reach midday at the same moment in Lecce as it does in Venice, or in Florence, or in Turin, because the sun moves from east to west. Midday arrives first in Venice, and significantly later in Turin, and for centuries the clocks in Venice were a good half-hour ahead of those in Turin. Every small village had its own peculiar 'hour'. A train station in Paris kept its own hour, a little behind the rest of the city, as a kind of courtesy towards travellers who were running late.[5]

In the nineteenth century the telegraph arrives, trains become commonplace and fast, and the problem arises of properly synchronizing clocks between one city and another. It is awkward to organize train timetables if each station marks time differently. It is in the United States that the first attempt is made to standardize time. Initially, it is proposed to fix a universal hour for the entire world. To call, for instance, 'twelve o'clock' the moment at which it is midday *in London*, so that midday would fall at 12.00 in London and around 18.00 in New York. The proposal is not well received, because people are attached to local time. In 1883 a compromise is reached with the idea of dividing the world into time zones, thereby standardizing time only within each zone. In this way, the discrepancy between twelve on the clock and local midday is limited to a maximum of about thirty minutes. The proposal is gradually accepted by the rest of the world and clocks begin to be synchronized between different cities.[6]

It can hardly be pure coincidence that, before

gaining a university position, the young Einstein worked in the Swiss Patent Office, dealing specifically with patents relating to the synchronization of clocks at railway stations. It was probably there that it dawned on him: the problem of synchronizing clocks was, ultimately, an insoluble one.

In other words, only a few years passed between the moment at which we agreed to synchronize clocks and the moment at which Einstein realized that it was impossible to do so exactly.

For millennia before clocks, our only regular way of measuring time had been the alternation of day and night. The rhythm of day followed by night also regulates the lives of plants and animals. Diurnal rhythms are ubiquitous in the natural world. They are essential to life, and it seems to me probable that they played a key role in the very origin of life on Earth, since an oscillation is required to set a mechanism in motion. Living organisms are full of clocks of various kinds – molecular, neuronal, chemical, hormonal – each of them more or less in tune with the others.[7] There are chemical mechanisms that keep to a twenty-four-hour rhythm even in the biochemistry of single cells.

The diurnal rhythm is an elementary source of our idea of time: night follows day; day follows night. We count the beats of this great clock: we count the days. In the ancient consciousness of humanity, time is, above all, this counting of days.

As well as the days, we then count the years and the

seasons, the cycles of the moon, the swings of a pendulum, the number of times that an hourglass is turned. This is the way in which we have traditionally conceived of time: counting the ways in which things change.

Aristotle is the first we are aware of to have asked himself the question 'What is time?', and he came to the following conclusion: time is the measurement of change. Things change continually. We call 'time' the measurement, the counting of this change.

Aristotle's idea is sound: time is what we refer to when we ask 'when?' 'After how much *time* will you return?' means 'When will you return?' The answer to the question 'when?' refers to something that happens. 'I'll return in three days' time' means that between departure and return the sun will have completed three circuits in the sky. It's as simple as that.

So if nothing changes, if nothing moves, does time therefore cease to pass?

Aristotle believed that it did. If nothing changes, time does not pass – because time is our way of situating ourselves in relation to the changing of things: the placing of ourselves in relation to the counting of days. Time is the measure of change:[8] if nothing changes, there is no time.

But what, then, is the time that I hear coursing in the silence? 'If it is dark and our bodily experience is nil,' Aristotle writes in his *Physics*, 'but some change is happening within the mind, we immediately suppose that some time has passed as well.'[9] In other words, even the time that we perceive flowing within us is the measure of a movement: a movement that is internal . . . If nothing moves, there is no time, because time is nothing but the registering of movement.

Newton instead assumes the exact opposite. In his magnum opus, the *Principia*, he writes:

> I do not define Time, Space, Place and Motion, as being well known to all. Only I must observe, that the common people conceive those quantities under no other notions but from the relation they bear to sensible objects. And thence arise certain prejudices, for the removing of which, it will be convenient to distinguish them into Absolute and Relative, True and Apparent, Mathematical and Common. [10]

In other words, Newton recognizes that a kind of 'time' exists that measures days and movements: the one treated by Aristotle (relative, apparent and common). But he also contends that, in addition to this, *another* time must exist: 'true' time that passes *regardless*, independently of things and of their changes. If all things remained motionless and even the movements of our souls were to be frozen, this time would continue to pass, according to Newton, unaffected and equal to itself: 'true' time. It's the exact opposite of what Aristotle writes.

'True' time, says Newton, is not directly accessible – only indirectly, through calculation. It is not the same

Aristotle: time is nothing other than the measurement of change.

Newton: there is a time that passes even when nothing changes.

as that given by days, because 'the natural days are truly unequal, though they are commonly consider'd as equal, and used for a measure of time: Astronomers correct this inequality that they may measure the celestial motions by a more accurate time.'[11]

So, who is right: Aristotle or Newton? Two of the most acute and profound investigators of nature that the world has ever seen are proposing two opposite ways of thinking about time. Two giants are pulling us in opposite directions.[12]

Time is only a way of measuring how things change, as Aristotle would have it – or should we be thinking that an absolute time exists that flows by itself, independently of things? The question we should really be asking is this: which of these two ways of thinking about time helps us to understand the world better? Which of the two conceptual schemes is more efficient?

For a few centuries, reason seemed to come down on the side of Newton. Newton's model, based on the idea of a time independent of things, has enabled the construction of modern physics – a physics that works incredibly well. And it assumes that time exists as an entity that runs in a way which is uniform and imperturbable. Newton writes equations containing the letter *t* for time which describe how things move *in time.*[13] What does this letter mean? Does *t* indicate time shaped by the longer hours of summer and the shorter ones of winter? Obviously not. It indicates time that is 'absolute, true and mathematical', assumed by Newton to run *independently of things that change or things that move.*

Clocks, for Newton, are devices that seek, albeit in a manner that is always imprecise, to follow this equal and uniform flowing of time. Newton writes that this 'absolute, true and mathematical' time is not perceptible. It must be deduced, through calculation and observation, from the regularity of phenomena. Newton's time is not the evidence given to us by our senses: it is an elegant intellectual construction. If, my dear, cultivated reader, the existence of this Newtonian concept of time which is independent of things seems to you simple and natural, it's because you encountered it at school. Because it has gradually become the way in which we all think about time. It has filtered through school textbooks throughout the world and ended up becoming our common way of understanding time. We have turned it into our common sense.

But the existence of a time that is uniform, independent of things and of their movement which *today* seems so natural to us is not an ancient intuition that is natural to humanity itself. It's an idea of Newton's.

The majority of philosophers have in fact responded negatively to this idea. In a still celebrated, furious counterblast, Leibniz defended the traditional thesis according to which time is only the order of events, arguing that there is no such thing as an autonomous time. Legend has it that Leibniz, whose name is still occasionally spelled with a 't' (Leibnitz), had deliberately dropped the letter from his name in accordance with his belief in the *nonexistence* of the absolute Newtonian time *t*.[14]

Before Newton, time for humanity was the way of counting how things changed. Before him, no one had thought it possible that a time independent of things could exist. Don't take your intuitions and ideas to be 'natural': they are often the products of the ideas of audacious thinkers who came before us.

But of these two giants, Aristotle and Newton, was it really Newton who was right? What exactly is this 'time' that he introduced, convincing the entire world that it exists: one that works brilliantly in his equations and yet is *not* the time that we perceive?

To get out from between these two giants, and in a strange way to reconcile them, a third was needed. Before getting to him, however, a brief digression on space is in order.

What is There, Where There is Nothing?

The two interpretations of time (the measure of 'when' with regard to events, as Aristotle wanted; the entity that runs even when nothing happens, according to Newton) can be repeated for space. Time is what we speak of when we ask 'when?' Space is what we speak of when we ask 'where?' If I ask, 'Where is the Coliseum?', one possible answer is: 'It's in Rome.' If I ask, 'Where are you?', a possible answer might be: 'At home.' To reply to the question 'Where is something?' means to indicate something else that is *around* that thing. If I say, 'In the Sahara,' you will visualize me surrounded by sand dunes.

Aristotle was the first to discuss in depth and with acuity the meaning of 'space', or 'place', and to arrive at a precise definition: the place of a thing is what surrounds that thing.[15]

As in the case of time, Newton suggests that we should think differently. The space defined by Aristotle, the enumeration of what surrounds each thing, is called 'relative, apparent and common' by Newton. He calls 'absolute, true and mathematical' space in itself, which exists even where there is nothing.

The difference between Aristotle and Newton is glaring. For Newton, between two things there may also be 'empty space'. For Aristotle, it is absurd to speak of 'empty' space, because space is only the spatial order of things. If there are no things – their extension, their contacts – there is no space. Newton imagines that things are situated in a 'space' that continues to exist, empty, even when divested of things. For Aristotle, this 'empty space' is nonsensical, because if two things do not touch it means that there is something else between them, and, if there is something, then this something is a thing, and therefore a thing that is there. It cannot be that there is 'nothing'.

For my part, I find it curious that both these ways of thinking about space originate from our everyday experience. The difference between them exists due to a quirky accident of the world in which we live: the lightness of air, the presence of which we only barely perceive. We can say: I see a table, a chair, a pen, the ceiling – and that between myself and the table *there is nothing*. Or we can say that between one and another of these things *there is air*. Sometimes we speak of air as if it were something, sometimes as if it were nothing. Sometimes as if it were there, sometimes as if it were not there. We are used to saying, 'This glass is empty' in order to say that it is full of air. We can consequently think of the world around us as 'almost empty', with just a few objects here and there, or alternatively as 'completely full', of air. In the end, Aristotle and Newton do not engage in profound metaphysics: they are only using these two different intuitive and ingenious ways of seeing the world around us – taking or not taking air into account – and transforming them into definitions of space.

Aristotle, always top of the class, wants to be precise: he does not say that the glass is empty; he says that it is full of air. And he remarks that, in our experience, there is never a place where 'there is nothing, not even air'. Newton, looking not so much for accuracy as for efficiency of the conceptual paradigm that needs to be constructed in order to describe the movement of things, thinks not about air but about objects. Air, after all, seems to have little effect on a falling stone. We can imagine that it is not even there.

As in the case of time, Newton's 'container space' may seem natural to us, but it is a recent idea that has spread due to the enormous influence of his thought. That which seems intuitive to us now is the result of scientific and philosophical elaborations in the past.

The Newtonian idea of 'empty space' seems to be confirmed when Torricelli demonstrates that it is possible to remove the air from a bottle. It soon becomes clear, however, that inside the bottle many physical entities remain: electric and magnetic fields, and a constant swarming of quantum particles. The existence of a complete void, without any physical entity except amorphous space, 'absolute, true and mathematical', remains a brilliant theoretical idea introduced by Newton to found his physics on, for there is no scientific, experimental evidence to support its existence. An ingenious hypothesis, perhaps the most profound insight achieved by one of the greatest scientists – but is it one that actually corresponds to the reality of things? Does Newton's space really exist? If it exists, is it really amorphous? Can a place exist where nothing exists?

The question is identical to the analogous one regarding time: does Newton's 'absolute, true and mathematical' time exist, flowing when nothing happens? If it exists, is it something altogether different from the things of this world? Is it so very independent from them?

The answer to all these questions lies in an unexpected synthesis of the apparently contradictory ideas held by these two giants. And, to accomplish this, it was necessary for a third giant to enter the dance.*

* I have been criticized for telling the history of science as if it were just the result of the ideas of a few brilliant minds rather than the painstaking work of generations. It's a fair enough criticism, and I apologize to the generations who have done and are doing the necessary work. My only excuse is that I am not attempting a detailed historical analysis or scientific methodology. I am only synthesizing a few crucial steps. Slow, technical, cultural and artistic advances made by innumerable workshops of painters and artisans were necessary before the Sistine Chapel was possible. But in the end, it was Michelangelo who painted it.

The Dance of the Three Giants

The synthesis between Aristotle's time and Newton's is the most valuable achievement made by Einstein. It is the crowning jewel of his thought.

The answer is that the time and space Newton had intuited the existence of, beyond tangible matter, do effectively *exist*. They are real. Time and space are real phenomena. But they are in no way absolute; they are not at all independent from what happens; they are not as different from the other substances of the world as Newton had imagined them to be. We can think of a great Newtonian canvas on which the story of the world is drawn. But this canvas is made of the same stuff that everything else in the world is made of, the same substance that constitutes stone, light and air: it is made of fields.

Physicists call 'fields' the substances which, to the best of our knowledge, constitute the weave of the physical reality of the world. Sometimes they may be given exotic names: 'Dirac fields' are the fabric of which tables and stars are made. The 'electromagnetic' field is the weave of which light is made, as well as the origin of the forces that make electric motors turn and the needle of a compass point north. But – here is the key point – there is also a 'gravitational' field: it is the origin of the force of gravity but it is also the texture that forms Newton's space and time, the fabric on which the rest of the world is drawn. Clocks are mechanisms that measure its extension. The meters used for measuring length are portions of matter which measure another aspect of its extension.

Spacetime is the gravitational field – and vice versa. It is something that exists by itself, as Newton intuited, even without matter. But it is not an entity that is different from the other things in the world – as Newton believed – it is a field like the others. More than a drawing on a canvas, the world is like a superimposition of canvases, of strata, where the gravitational field is only one among others. Just like the others, it is neither absolute nor uniform, nor is it fixed: it flexes, stretches and jostles with the others, pushing and pulling against them. Equations describe the reciprocal influences that all the fields have on each other, and spacetime is one of these fields.*

* The route by which Einstein arrived at this conclusion was a long one: it didn't end with the writing of the equations of the field in 1915 but continued in tortuous efforts to understand its physical significance, causing him in the process to change his ideas repeatedly. He was confused in particular regarding the existence of solutions without matter, and by whether gravitational waves were real or not. He achieves definitive clarity only in his last writings and, in particular, in the fifth appendix, 'Relativity and the Problem of Space', which was added to the fifth edition of *Relativity: The Special and General Theory* (Methuen, London, 1954). This appendix can be read at http://www.relativitybook.com/resources/Einstein_space.html. For copyright reasons, this appendix is

The gravitational field can also be smooth and flat, like a straight surface, and this is the version that Newton described. If we measure it with a meter, we discover that the Euclidean geometry that we learned at school applies. But the field can also undulate, in what we call gravitational waves. It can contract and expand.

Remember the clocks in Chapter 1 that slow down in the vicinity of a mass? They slow down because there is, in a precise sense, 'less' gravitational field there. There is less time there.

The canvas formed by the gravitational field is like a vast elastic sheet that can be pulled and stretched. Its stretching and bending is the origin of the force of gravity, of things falling, and provides a better explanation of this than the old Newtonian theory of gravity.

Look again at the figure in Chapter 1 that illustrates how more time passes above than below, but imagine now that the piece of paper on which the diagram is drawn is elastic; imagine stretching it so that the time in the mountains actually becomes elongated. You will obtain something like the image opposite, which represents space (the height, on the vertical axis) and time (on the horizontal) – but, now,

---

not included in most editions of the book. A more in-depth discussion can be found in Chapter 2 of my *Quantum Gravity* (Cambridge University Press, Cambridge, 2004).

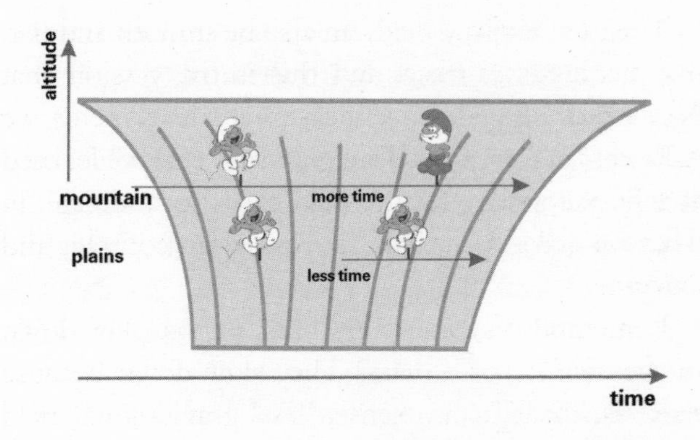

the 'longer' time in the mountains effectively corresponds to a greater length of time.

This image illustrates what physicists call 'curved' spacetime. 'Curved' because it is distorted: distances are stretched and contracted, just like the elastic sheet when it is pulled. This is why the light cones were inclined in the diagrams in Chapter 3.

Time thus becomes part of a complicated geometry woven together with the geometry of space. This is the synthesis that Einstein found between Aristotle's conception of time and Newton's. With a tremendous beat of his wings, Einstein understands that Aristotle and Newton are *both* right. Newton is right in intuiting that something else exists in addition to the simple things that we see moving and changing. True and mathematical Newtonian time exists; it is a real entity; it is the gravitational field, the elastic sheet, the curved spacetime in the diagram. But Newton is wrong in assuming that this time is independent from things – and that it passes regularly, imperturbably, separately from everything else.

For his part, Aristotle is right to say that 'when' and 'where' are always located in relation to something. But this something can also be just the field, the spatio-temporal entity of Einstein, because this is a dynamic and concrete entity, like all those in reference to which, as Aristotle rightly observed, we are capable of locating ourselves.

All this is perfectly coherent, and Einstein's equations describing the distortions of the gravitational field and its effects on clocks and meters have been repeatedly verified for more than a century. But our idea of time has lost another of its constituent parts: its supposed independence from the rest of the world.

The three-handed dance of these intellectual giants – Aristotle, Newton and Einstein – has guided us to a deeper understanding of time and of space. There is a structure of reality that is the gravitational field; it is not separate from the rest of physics, nor is it the stage across which the world passes. It is a dynamic component of the great dance of the world, similar to all the others, interacting with the others, determining the rhythm of those things that we call meters and clocks and the rhythm of all physical phenomena.

Success, as ever, is destined to be short-lived – even great success. Einstein writes the equations of the gravitational field in 1915, and barely a year later it is Einstein himself who observes that this cannot be the last word on the nature of time and space, because of the existence of quantum mechanics. The gravitational field, like all physical things, must necessarily have quantum properties.

5. Quanta of Time

> There's an amphora
> of old wine in the house
> of nine years' vintage.
> There is in the garden, Phyllis,
> laurel for braiding crowns
> and much ivy . . .
> I invite you to celebrate
> this day in mid-April –
> a festive one for me,
> dearer than my own birthday. (IV, 11)

The strange landscape of the physics of relativity that I have described so far becomes even more alien when we consider quanta and the quantum properties of space and of time.

The discipline which studies these is called 'quantum gravity', and this is my own field of research.[1] There is not yet a theory of quantum gravity that has been generally accepted by the scientific community, or obtained experimental support. My scientific life has been largely dedicated to contributing to the construction of a possible solution to the problem: *loop* quantum gravity, or *loop theory*. Not everyone is betting on this turning out to be the right solution. Friends who work on string theory, for instance, are following different paths, and the battle to establish who is right is still raging. This is good – science also grows thanks to fierce debates: sooner or later, it will become clear which theory is correct, and perhaps we will not have long to wait.

Divergences of opinion regarding the nature of time, however, have diminished in the last few years, and many conclusions have become reasonably clear to most. What has been clarified is that the residual temporal scaffolding of general relativity, illustrated in the previous chapter, also falls away if we take quanta into account.

Universal time has shattered into a myriad of proper times but, if we factor in the quanta, we must accept the idea that each of these times, in turn, 'fluctuates' and is dispersed as in a cloud – and can have only certain values and not others . . . They can no longer manage to form the spatio-temporal sheet outlined in the previous chapters.

The three fundamental discoveries that quantum mechanics has led to are these: granularity, indeterminacy and the relational aspect of physical variables. Each one of these demolishes further the little that was left of our idea of time. Let's consider them one by one.

Granularity

The time measured by a clock is 'quantified', that is to say, it acquires only certain values and not others. It is as if time were granular rather than continuous.

Granularity is the most characteristic feature of quantum mechanics, which takes its name from this: 'quanta' are elementary grains. A minimum scale exists for all phenomena.[2] For the gravitational field, this is called the 'Planck scale'. Minimum time is called 'Planck time'. Its value can be easily estimated by combining the constants that characterize phenomena subject to relativity, gravity and quantum mechanics.[3] Together, these determine the time to 10^{-44} seconds: a hundred millionth of a trillionth of a trillionth of a trillionth of a second. This is Planck time: at this extremely minuscule level, quantum effects on time become manifest.

Planck's time is small, much smaller than any actual clock is today capable of measuring. It is so extremely small that we should not be astounded to discover that 'down there', at such a minute scale, the notion of time is no longer valid. Why *should* it be? Nothing is valid always and everywhere. Sooner or later, we always come across something that is new.

The 'quantization' of time implies that almost all values of time *t do not exist.* If we could measure the duration of an interval with the most precise clock imaginable, we should find that the time measured takes only certain discrete, special values. It is not possible to think of duration as continuous. We must think of it as discontinuous: not as something which flows uniformly but as something which in a certain sense jumps, kangaroo-like, from one value to another.

In other words, a *minimum* interval of time exists. Below this, the notion of time does not exist – even in its most basic meaning.

Perhaps the rivers of ink which have been expended discussing the nature of the 'continuous' over the centuries, from Aristotle to Heidegger, have been wasted. Continuity is only a mathematical technique for approximating very finely grained things. The world is subtly discrete, not continuous. The good Lord has not drawn the world with continuous lines: with a light hand, he has sketched it in dots, like Seurat.

Granularity is ubiquitous in nature: light is made of photons, the particles of light. The energy of electrons in atoms can acquire only certain values and not others. The purest air is granular, and so, too, is the densest matter. Once it is understood that Newton's space and time are physical entities like all others, it is natural to suppose that they are also granular. Theory confirms this idea: loop quantum gravity predicts that elementary temporal leaps are small, but finite.

The idea that time could be granular, that there could be minimal intervals of time, is not new. It is defended in the seventh century by Isidore of Seville in his *Etymologiae*, and in the following century by the

Venerable Bede in a work suggestively entitled *De Divisionibus Temporum*: 'On the Divisions of Time'. In the twelfth century the great philosopher Maimonides writes: 'Time is composed of atoms, that is to say of many parts that cannot be further subdivided, on account of their short duration.'[4] The idea probably dates back even further: the loss of the original texts of Democritus prevents us from knowing whether it was already present in classical Greek atomism.[5] Abstract thought can anticipate by centuries hypotheses that find a use – or confirmation – in scientific inquiry.

The spatial sister of *Planck time* is *Planck length*: the minimum limit below which the notion of length becomes meaningless. Planck length is around 10^{-33} centimetres: a millionth of a billionth of a billionth of a billionth of a millimetre. As a young man, at university, I fell in love with the problem of what happens at this extremely small scale. I painted a large sheet with, at its centre, in red, a glittering:

I hung it up in my bedroom in Bologna and decided that my goal would be to try to understand what happens down there, at the very tiny scale where time and space cease to be what they are – all the way to the elementary quanta of space and time. Then I spent the rest of my life actually trying to achieve this.

Quantum Superpositions of Times

The second discovery made by quantum mechanics is indeterminacy: it is not possible to predict exactly, for instance, where an electron will appear tomorrow. Between one appearance and another, the electron has no precise position,[6] as if it were dispersed in a cloud of probability. In the jargon of physicists, we say that it is in a 'superposition' of positions.

Spacetime is a physical object like an electron. It, too, fluctuates. It, too, can be in a 'superposition' of different configurations. The illustration of stretched time at the end of Chapter 4, for example – if we take quantum mechanics into account – should be imagined as a blurred superposition of different spacetimes, more or less as shown here:

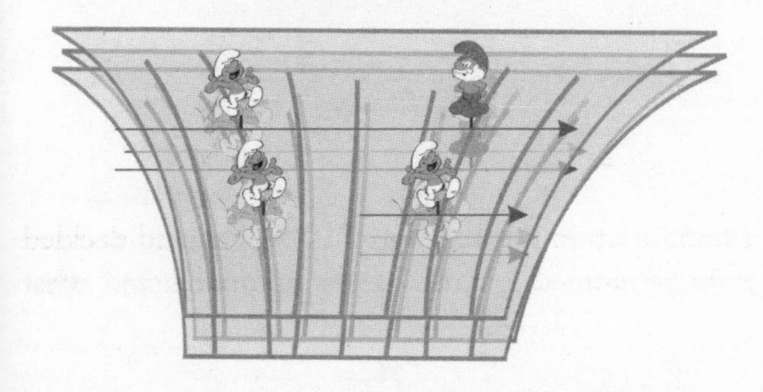

Similarly, the structure of light cones fluctuates at all points that distinguish between past, present and future:

Even the distinction between present, past and future thus becomes fluctuating, indeterminate. Just as a particle may be diffused in space, so, too, the differences between past and future may fluctuate: an event may be both before and after another one.

Relations

'Fluctuation' does not mean that what happens is *never* determined. It means that it is determined only at certain moments, and in an unpredictable way.

Indeterminacy is resolved when a quantity interacts with something else.*

In the interaction, an electron materializes at a certain point. For example, it collides with a screen, is captured by a particle detector, or collides with a photon – thus acquiring a concrete position.

But there is a strange aspect to this materialization of the electron: the electron is concrete only *in relation to* the other physical objects it is interacting with. With regard to all the others, the effect of the interaction is only to spread the contagion of indeterminacy. Concreteness occurs only in relation to a physical system: this, I believe, is the most radical discovery made by quantum mechanics.†

When an electron collides with an object – the screen of an old television set with a cathode tube, for example – the cloud of probability with which we conceived of it 'collapses' and the electron materializes at a point on the screen, producing one of the

* The technical term for interaction used in this context, 'measure', is misleading because it seems to imply that in order to create reality we need an experimental physicist in a white coat.

† I'm making use here of the relational interpretation of quantum mechanics,[7] which is the one I myself find least implausible. The observations which follow, in particular the loss of the classical spacetime satisfying Einstein's equations, remain invalid in every other interpretation that I know of.

luminous dots that goes into the making of a TV image. But it is only in relation to the screen that this happens. In relation to another object, the electron and screen are now together in a superposition of configurations, and it is only at the moment of further interaction with a third object that their shared cloud of probability 'collapses' and materializes in a particular configuration – and so on.

It is hard to take in the idea that an electron behaves in such a bizarre way. It is even more difficult to digest that this is also the way time and space behave. And yet, according to all the evidence, this is the way the quantum world works: the world which we inhabit.

The physical substratum that determines duration and physical intervals – the gravitational field – does not only have a dynamic influenced by masses; it is also a quantum entity that does not have determined values until it interacts with something else. When it does, the durations are granular and determinate only for that something with which it interacts; they remain indeterminate for the rest of the universe.

Time has loosened into a network of relations that no longer holds together as a coherent canvas. The picture of spacetimes (in the plural) fluctuating, superimposed one above the other, materializing at certain times with respect to particular objects, provides us with a very vague vision. But it is the best that we have for the fine granularity of the world. We are peering into the world of quantum gravity.

Let me reprise the long dive into the depths made in the first part of this book. There is no single time: there is a different duration for every trajectory; and time passes at different rhythms according to place and according to speed. It is not directional: the difference between past and future does not exist in the elementary equations of the world; its orientation is merely a contingent aspect that appears when we look at things and neglect the details. In this blurred view, the past of the universe was in a curiously 'particular' state. The notion of the 'present' does not work: in the vast universe there is nothing that we can reasonably call 'present'. The substratum that determines the duration of time is not an independent entity, different from the others that make up the world; it is an aspect of a dynamic field. It jumps, fluctuates, materializes only by interacting, and is not to be found beneath a minimum scale . . . So, after all this, what is left of time?

> You got to deep-six your wristwatch, you got to try and understand,
>
> The time it seems to capture is just the movement of its hands . . .[8]

Let's enter the world without time.

PART TWO
The World without Time

6. The World is Made of Events, not Things

> O gentlemen, the time of life is short [. . .]
> And if we live, we live to tread on kings.
> Shakespeare, *Henry IV, Part 1* (V, ii, 81, 85)

When Robespierre freed France from monarchy, Europe's *ancien regime* feared that the end of civilization itself was nigh. When the young seek to liberate themselves from an old order of things, the old are afraid that all will founder. But Europe was able to survive perfectly well, even without the King of France. The world will go on turning, even without King Time.

There is, nevertheless, an aspect of time that has survived the demolition inflicted on it by nineteenth- and twentieth-century physics. Divested of the trappings with which Newtonian theory had draped it, and to which we had become so accustomed, it now shines out with greater clarity: the world is nothing but change.

None of the pieces that time has lost (singularity, direction, independence, the present, continuity) puts into question the fact that the world is a network of *events*. On the one hand, there was time, with its many determinations; on the other, the simple fact that nothing *is*: that things happen instead.

The absence of the quantity 'time' in the fundamental equations does not imply a world that is frozen and immobile. On the contrary, it implies a world in which change is ubiquitous, without being ordered by Father Time; without innumerable events being necessarily distributed in good order, or along the single Newtonian timeline, or according to Einstein's elegant geometry. The events of the world do not form an orderly queue, like the English. They crowd around chaotically, like Italians.

They are events, indeed: change, happening. This happening is diffuse, scattered, disorderly. But it is happening; it is not stasis. Clocks that run at different speeds do not mark a single time, but the hands on each clock change in relation to the others. The fundamental equations do not include a time variable, but they do include variables that change in relation to each other. Time, as Aristotle suggested, is the measure of change; different variables can be chosen to measure that change, and none of these has *all* the characteristics of time as we experience it. But this does not alter the fact that the world is in a ceaseless process of change.

The entire evolution of science would suggest that the best grammar for thinking about the world is that of change, not of permanence. Not of being, but of becoming.

We can think of the world as made up of *things*. Of *substances*. Of *entities*. Of something that *is*. Or we can think of it as made up of *events*. Of *happenings*. Of *processes*. Of something that *occurs*. Something that does not last, and that undergoes continual transformation, that is not permanent in time. The destruction of the notion of time in fundamental physics is the crumbling of the first of these two perspectives, not of the second. It is the realization of the ubiquity of impermanence, not of stasis in a motionless time.

Thinking of the world as a collection of events, of processes, is the way that allows us to better grasp, comprehend and describe it. It is the only way that is compatible with relativity. The world is not a collection of things, it is a collection of events.

The difference between things and events is that *things* persist in time; *events* have a limited duration. A stone is a prototypical 'thing': we can ask ourselves where it will be tomorrow. Conversely, a kiss is an 'event'. It makes no sense to ask where the kiss will be tomorrow. The world is made up of networks of kisses, not of stones.

The basic units in terms of which we comprehend the world are not located in some specific point in space. They are – if they are at all – in a *where* but also in a *when*. They are spatially but also temporally delimited: they are events.

On closer inspection, in fact, even the things that are most 'thing-like' are nothing more than long events. The hardest stone, in the light of what we have learned from chemistry, from physics, from mineralogy, from geology, from psychology, is in reality a complex vibration of quantum fields, a momentary interaction of forces, a process that for a brief moment manages to keep its shape, to hold itself in equilibrium before disintegrating again into dust, a brief chapter in the history of interactions between the elements of the planet, a trace of Neolithic humanity, a weapon used by a gang of kids, an example in a book about time, a metaphor for an ontology, a part of a segmentation of the world that depends more on how our bodies are structured to perceive than on the object of perception – and, gradually, an intricate knot in that cosmic game of mirrors that constitutes reality. The world is not so much made of stones as of fleeting sounds, or of waves moving through the sea.

If the world were, however, made of things, what would these things be? The atoms, which we have discovered to be made up in turn of smaller particles? The elementary particles, which, as we have discovered, are nothing other than the ephemeral agitations of a field? The quantum fields, which we have found to be little more than codes of a language with which to speak of interactions and events? We cannot think of the *physical* world as if it were made of things, of entities. It simply doesn't work.

What works instead is thinking about the world as a network of events. Simple events, and more complex events that can be disassembled into combinations of simpler ones. A few examples: a war is not a thing, it's a sequence of events. A storm is not a thing, it's a collection of occurrences. A cloud above a mountain is not a thing, it is the condensation of humidity in the air that the wind blows over the mountain. A wave is not a thing, it is a movement of water, and the water that forms it is always different. A family is not a thing, it is a collection of relations, occurrences, feelings. And a human being? Of course it's not a thing; like the cloud above the mountain, it's a complex process which food, information, light, words and so on enter and exit . . . A knot of knots in a network of social relations, in a network of chemical processes, in a network of emotions exchanged with its own kind.

For a long time, we have tried to understand the world in terms of some primary *substance*. Perhaps physics, more than any other discipline, has pursued this primary substance. But the more we have studied it, the less the world seems comprehensible in terms of something that *is*. It seems to be a lot more intelligible in terms of relations between events.

The words of Anaximander quoted in the first chapter of this book invited us to think of the world 'according to the order of time'. If we do not assume *a priori* that we know *what* the order of time is, if we do not, that is, presuppose that it is the linear and universal order that we are accustomed to, Anaximander's exhortation remains valid: we understand the world by studying change, not by studying things.

Those who have neglected this good advice have paid a heavy price for it. Two of the greats who fell into this error were Plato and Kepler, both curiously seduced by the same mathematics.

In the *Timaeus*, Plato has the excellent idea of attempting to translate into mathematics the physics insights gained by atomists such as Democritus. But he goes about it the wrong way: he tries to write the mathematics of the *shape* of atoms, rather than the mathematics of their *movements*. He allows himself to be fascinated by a mathematical theorem which establishes that there are five – and *only* five – regular polyhedra:

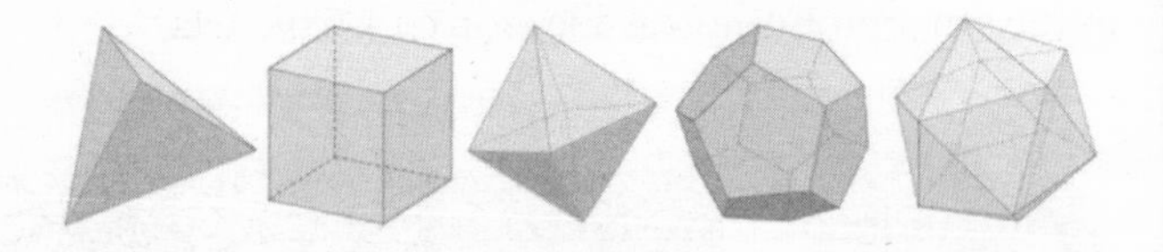

And he attempts to advance the audacious hypothesis that these are the actual shapes of the atoms of the five elementary substances that in antiquity were thought to form everything: earth, water, air, fire and the quintessence of which the heavens are made. A beautiful idea. But completely mistaken. The error lies in seeking to understand the world in terms of things rather than events. It lies in ignoring change. The physics and astronomy that will work, from Ptolemy to Galileo, from Newton to Schrödinger, will be mathematical descriptions of precisely how things *change*, not of how they *are*. They will be about events, not things. The shapes of atoms will be eventually understood only with solutions to Schrödinger's equations describing how the electrons in atoms *move*. Events again, not things.

Centuries later, before reaching the momentous results obtained in his maturity, the young Kepler falls into the same error. He asks himself what determines the size of the orbits of planets and allows himself to be seduced by the same theorem that had bewitched Plato (it's a beautiful one, no doubt about that). Kepler assumes that the regular polyhedra determine the size of the orbits of the planets: if they are placed one inside the other with spheres between them, the radii of these spheres will be in the same proportion as the radii of the planets.

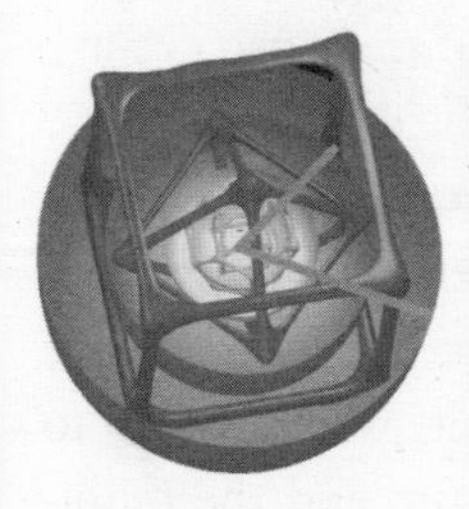

A nice idea, but completely wrong-headed. Once again, it lacks dynamics. When Kepler moves on, later in life, to tackle the question of how planets *move*, the gates of the heavens open.

We therefore describe the world as it happens, not as it is. Newton's mechanics, Maxwell's equations, quantum mechanics, and so on, tell us how *events happen*, not how *things are*. We understand biology by studying how living beings *evolve* and *live*. We understand psychology (a little, not much) by studying how we interact with each other, how we think . . . We understand the world in its becoming, not in its being.

'Things' in themselves are only events that for a while are monotonous.[1]

But only before returning to dust. Because sooner or later, obviously, everything returns to dust.

The absence of time does not mean, therefore, that everything is frozen and unmoving. It means that the incessant happening that wearies the world is not ordered along a timeline, is not measured by a gigantic tick-tocking. It does not even form a four-dimensional geometry. It is a boundless and disorderly network of quantum events. The world is more like Naples than Singapore.

If by 'time' we mean nothing more than happening, then everything is time. There is *only* that which exists in time.

7. The Inadequacy of Grammar

Gone is the whiteness
of snow –
green returns
in the grass of the fields,
in the canopies of trees,
and the airy grace of spring
is with us again.
Thus time revolves,
the passing hour that steals
the light
brings a message:
immortality, for us, is impossible.
Warm winds will be followed by cold. (IV, 7)

Usually, we call 'real' the things that exist *now*, in the present. Not those which existed once, or may do so in the future. We say that things in the past or the future 'were' real or 'will be' real, but we do not say they 'are' real.

Philosophers call 'presentism' the idea that only the present is real, that the past and the future are not – and that reality evolves from one present to another, successive one.

This way of thinking no longer works, however, if

the 'present' is not defined globally, if it is defined only in our vicinity, in an approximate way. If the present that is far away from here is not defined, what 'is real' in the universe?

Diagrams such as the ones that we have seen in previous chapters depict *an entire evolution* of spacetime with a single image: they do not represent a single time but all times together:

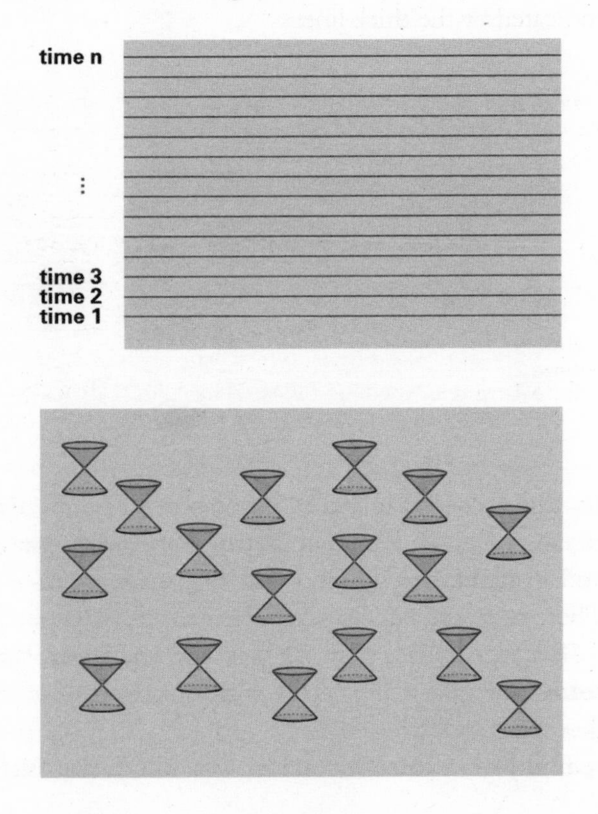

They are like a sequence of photographs of a man running, or a book containing a story that develops over many years. They are a schematic representation of a possible *history* of the world, not of one of its single, instantaneous states.

The diagram below illustrates how we used to think of the temporal structure of the world *before* Einstein. The set of real events *now*, at a given time, is indicated by the thick line:

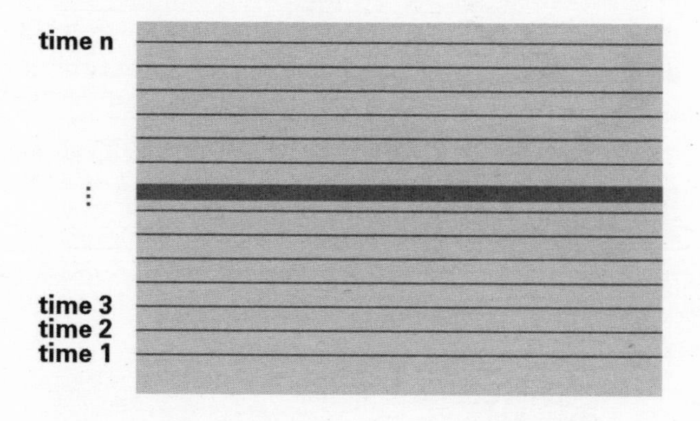

But the second diagram above provides a better representation of the temporal structure of the world, and in it there is nothing that resembles a present. There is no present. So what is real *now*?

Twentieth-century physics shows, in a way that seems unequivocal to me, that our world is not described well by *presentism*: an objective global present does not exist. The most we can speak of is a present relative to a moving observer. But then, what is real for me is different from that which is real for you, despite the fact that we would like to use the expression 'real' – in an objective sense – as much as possible. Therefore, the world should not be thought of as a succession of presents.[1]

What alternatives do we have?

Philosophers call 'eternalism' the idea that flow and change are illusory: present, past and future are all equally real and equally existent. Eternalism is the idea that the whole of spacetime, as outlined in the above diagrams, exists all together in its entirety without anything changing. Nothing really flows.[2]

Those who defend this way of thinking about reality – eternalism – frequently cite Einstein, who in a famous letter writes:

> People like us who believe in physics, know that the distinction between past, present and future is only a stubbornly persistent illusion.[3]

This idea has come to be called the *block universe*: the idea that it is necessary to think of the history of the universe as a single block, all equally real, and that the passage from one moment of time to the next is illusory.

And so is this – eternalism, the block universe – the only way left for us to conceive of the world? Must we think of the world with past, present and future like a single present, all existing in the same way? That nothing changes, and that everything is motionless? Is change only an illusion?

No, I really don't think so.

The fact that we cannot arrange the universe like a single orderly sequence of times does not mean that nothing changes. It means that changes are not arranged in a single orderly succession: the temporal structure of the world is more complex than a simple single linear succession of instants. This does not mean that it is non-existent or illusory.[4]

The distinction between past, present and future is not an illusion. It is the temporal structure of the world. But the temporal structure of the world is not that of presentism. The temporal relations between events are more complex than we previously thought, but they do not cease to exist on account of this. The relations of filiation do not establish a global order, but this does not make them illusory. If we are not all in single file, it does not follow that there are no relations between us. Change, what happens – this is not an illusion. What we have discovered is that it does not follow a global order.[5]

Let's return to the question with which we began: what 'is real'? What 'exists'?

The answer is that this is a badly put question, signifying everything and nothing. Because the adjective 'real' is ambiguous, it has a thousand meanings. The verb 'to exist' has even more. To the question 'Does a puppet whose nose grows when he lies exist?', it is possible to reply: 'Of course he exists! It's Pinocchio!'; or: 'No, he doesn't, he's only part of a fantasy dreamed up by Collodi.'

Both answers are correct, because they are using different meanings of the verb 'to exist'.

There are so many different usages of the verb; different ways in which we can say that a thing exists: a law, a stone, a nation, a war, a character in a play, the god (or gods) of a religion to which we do not belong, the God of the religion to which we do belong, a great love, a number . . . Each one of these entities 'exists' and 'is real' in a sense different from all the others. We can ask ourselves in what sense something exists or not (Pinocchio exists as a literary character but not as far as any Italian registry office is concerned), or if a thing exists in a determined way (does a rule exist preventing you from 'castling' in chess, if you have already moved the castle?). To ask oneself in general 'What exists?' or 'What is real?' means only to ask how you would like to use a verb and an adjective.[6] It's a grammatical question, not a question about nature.

Nature, for its part, is what it is – and we discover it very gradually. If our grammar and our intuition do not readily adapt to what we discover, well, too bad: we must seek to adapt them.

The grammar of many modern languages conjugates verbs in the 'present', 'past' and 'future' tenses. It is not well adapted for speaking about the real temporal structure of reality, which is more complex. Grammar developed from our limited experience, before we became aware of its imprecision when it came to grasping the rich structure of the world.

What confuses us when we seek to make sense of the discovery that no objective universal present exists is only the fact that our grammar is organized around an absolute distinction – 'past/present/future' – that is only partially apt, here in our immediate vicinity. The structure of reality is not the one that this grammar presupposes. We say that an event 'is', or 'has been', or 'will be'. We do not have a grammar adapted to say that an event 'has been' in relation to me but 'is' in relation to you.

We must not allow ourselves to be confused by an inadequate grammar. There is a text from the world of antiquity which refers to the spherical shape of the Earth in the following way:

> For those standing below, things above are below, while things below are above . . . and this is the case around the entire earth.[7]

On first reading, the phrase is a muddle, a contradiction in terms. How is it possible that 'things above are below, while things below are above'? It makes no sense. It is comparable to the sinister 'Fair is foul and foul is fair' in *Macbeth*. But if we reread it bearing in mind the shape and the physics of the Earth, the phrase becomes clear: its author is saying that for those who live at the

Antipodes (in Australia) the direction 'upwards' is the same as 'downwards' for those who are in Europe. He is saying, that is, that the direction 'above' changes from one place to another on the Earth. He means that what is above *with respect to Sydney* is below *with respect to us.* The author of this text, written two thousand years ago, is struggling to adapt his language and his intuition to a new discovery: the fact that the Earth is a sphere, and that 'up' and 'down' have a meaning that *changes* between here and there. The terms do not have, as previously thought, a single and universal meaning.

We are in the same situation. We are struggling to adapt our language and our intuition to a new discovery: the fact that 'past' and 'future' do not have a universal meaning. Instead, they have a meaning which changes between here and there. That's all there is to it.

In the world, there is change, there is a temporal structure of relations between events that is anything but illusory. It is not a global happening. It is a local and complex one which is not amenable to being described in terms of a single global order.

And what about Einstein's phrase 'the distinction between past, present and future is only a stubbornly persistent illusion'? Does it not seem to say that he thought the opposite? Even if this were the case, I am not sure that because Einstein has penned some phrase or other we should treat it as the utterance of an oracle. Einstein changed his mind many times on fundamental questions, and it is possible to find numerous erroneous phrases of his that contradict each other.[8] But, in this instance, things are perhaps much simpler. Or more profound.

Einstein coins this phrase when his friend Michele Besso dies. Michele has been his dearest friend, the companion of his thinking and discussions since his days at the University of Zurich. The letter in which Einstein writes the phrase is not directed at physicists or philosophers. It is addressed to Michele's family, in particular to his sister. The sentence that comes before it reads:

> Now he [Michele] has departed from this strange world a little ahead of me. That means nothing . . .

It is not a letter written to pontificate about the structure of the world: it's a letter written to console a grieving sister. A gentle letter, alluding to the spiritual bond between Michele and Albert. A letter in which Einstein also confronts his own suffering at the loss of his lifelong friend; and in which, evidently, he is thinking about his own approaching death. A deeply emotional letter, in which the illusoriness and the heart-rending irrelevance to which he alludes do not refer to time as understood by physicists. They are prompted by the experience of life itself. Fragile, brief, full of illusions. It's a phrase that speaks of things that lie deeper than the *physical* nature of time.

Einstein died on 18 April 1955, one month and three days after the death of his friend.

8. Dynamics as Relation

> Sooner or later
> the exact measurement of our time
> will resume –
> and we'll be on the ship that's bound
> for the bitterest shore. (II, 9)

How does one describe a world in which everything occurs but there is no time variable? In which there is no common time and no privileged direction in which change occurs?

In the simplest way, the same way that we had thought about the world until Newton convinced us all that a time variable was indispensable.

To describe the world, the time variable is not required. What is required are variables that actually describe it: quantities that we can perceive, observe and eventually measure. The length of a road, the height of a tree, the temperature of a forehead, the weight of a piece of bread, the colour of the sky, the number of stars in the celestial vault, the elasticity of a piece of bamboo, the speed of a train, the pressure of a hand on a shoulder, the pain of a loss, the position of the hands on a clock, the height of the sun in the sky . . . These are the terms in which we describe the world. Quantities and properties that we see continuously *changing*. In these changes there are regularities: a stone falls faster than a feather. Sun and moon circle after each other in the sky, passing by each other once a month . . . Among these quantities there are some that we see changing regularly with respect to others: the number of days, the phases of the moon, the height of the sun on the horizon, the position of the hands of a clock. It is useful to employ *these* as points of reference: let's meet three days after the next full moon, when the sun is at its highest in the sky. I'll see you tomorrow, when the clock shows 4:35. If we find a sufficient number of variables that remain synchronized enough in relation to each other, it is convenient to use them in order to speak of *when*.

There is no need in any of this to choose a privileged variable and call it 'time'. What we need, if we want to do science, is a theory that tells us how the variables change with respect to each other. That is to say, how one changes when others change. The fundamental theory of the world must be constructed in this way; it does not need a time variable: it needs to tell us only how the things that we see in the world vary with respect to each other. That is to say, what the relations may be between these variables.[1]

The fundamental equations of quantum gravity are effectively formulated like this: they do not have a time variable, and they describe the world by indicating the possible relations between variable quantities.[2]

In 1967 an equation accounting for quantum gravity was written for the first time without any time variable. This equation was discovered by two American physicists – Bryce DeWitt and John Wheeler – and today it's known as the Wheeler–DeWitt equation.[3]

At first no one could understand the significance of an equation without a time variable, perhaps not even Wheeler and DeWitt themselves. (Wheeler: 'Explain time? Not without explaining existence! Explain existence? Not without explaining time! To uncover the deep and hidden connection between time and existence . . . is a task for the future.')[4] The issue was discussed at great length; there were conferences, debates; rivers of ink flowed.[5] I think that the dust has now settled and things have become much clearer. There is nothing mysterious about the absence of time in the fundamental equation of quantum gravity. It is only the consequence of the fact that, at the fundamental level, no special variable exists.

The theory does not describe how things evolve *in time*. The theory describes how things change *one in respect to the others*,[6] how things happen in the world in relation to each other. That's all there is to it.

Bryce and John left us some years ago. I knew them both and had great admiration and respect for them. In my study at the University of Marseille I have hanging on the wall a letter that John Wheeler wrote to me when he became aware of my first work on quantum gravity. Every so often I re-read it with a mixture of pride and nostalgia. I would like to have asked him more, during the handful of meetings that we had together. The last time I went to see him in Princeton we took a long walk together. He spoke to me with the soft voice of an old man. I could not make out much of what he said but did not dare to ask him too frequently to repeat what he was saying. Now he is no longer with us. I can't question him any more, or tell him what I think. I can no longer tell him that it seems to me that his ideas are correct, and that they have guided me throughout a lifetime of research. I can no longer tell him I believe that he was the first to come close to the heart of the mystery of quantum gravity. Because he is no longer here – here and now. This is time for us. Memory and nostalgia. The pain of absence.

But it isn't absence that causes sorrow. It is affection and love. Without affection, without love, such absences would cause us no pain. For this reason, even the pain caused by absence is, in the end, something good and even beautiful, because it feeds on that which gives meaning to life.

I met Bryce in London, on the first occasion that I went to seek out a group working on quantum gravity. I was a young novice, fascinated by this arcane subject that no one in Italy was working on; he was its grand guru. I had gone to Imperial College to meet up with Chris Isham, and when I arrived I was told that he was on the top-floor terrace. Sitting together at a small table up there I found Chris Isham, Karel Kuchar and Bryce DeWitt – the three main authors whose ideas I had been studying in recent years. I remember vividly the intense impression that I had on seeing them there through the glass, calmly discussing among themselves. I did not dare to interrupt. They seemed to me like three great Zen masters exchanging unfathomable truths between enigmatic smiles.

They were probably only deciding where to go for dinner. Revisiting and reflecting on the episode, I realize that at the time they were younger than I am now. This, too, is time: a strange shifting of perspective. Shortly before he died, Bryce gave a long interview in Italy, subsequently published in a small book.[7] And it was only then I became aware that he'd followed my work much more closely, and with more sympathy, than I had ever suspected from our conversations, in which he was more prone to express criticism than encouragement.

John and Bryce were my spiritual fathers. Thirsting, I found in their ideas fresh, clear water to drink. So, thank you, John; thanks, Bryce. As human beings, we live by emotions and thoughts. We exchange them when we are in the same place at the same time, talking to each other, looking into each other's eyes, brushing against each other's skin. We are nourished by this network of encounters and exchanges. But, in reality, we do not need to be in the same place and time to have such exchanges. Thoughts and emotions that create bonds of attachment between us have no difficulty in crossing seas and decades, sometimes even centuries, tied to thin sheets of paper or dancing between the microchips of a computer. We are part of a network that goes far beyond the few days of our lives and the few square metres that we tread. This book is also a part of that weave . . .

But I have digressed and lost my thread. Nostalgia for John and Bryce has caused me to deviate from my path. All I had intended to say in this chapter is that they had discovered the extremely simple structure of the equation that describes the dynamics of the world. It describes possible events and the correlations between them, and nothing else.

It's the elementary form of the mechanics of the world, and it does not need to mention 'time'. The world without a time variable is not a complicated one. It's a net of interconnected events, where the variables in play adhere to probabilistic rules which, incredibly, we know for a good part how to write. And it's a clear world, windswept and full of beauty as the crests of mountains; as beautiful as the cracked lips of adolescents.

Elementary Quantum Events and Spin Networks

The equations of loop quantum gravity[8] on which I work are a modern version of the theory of Wheeler and DeWitt. There is no time variable in these equations.

The variables of the theory describe the fields that form matter, photons, electrons, other components of atoms and the gravitational field – all on the same level. Loop theory is not a 'unified theory of everything'. It doesn't even begin to claim that it's the ultimate theory of science. It's a theory made up of coherent but distinct parts. It seeks to be 'only' a *coherent* description of the world as we understand it so far.

The fields manifest themselves in granular form: elementary particles, photons and quanta of gravity – or rather 'quanta of space'. These elementary grains do not exist immersed in space; rather, they themselves form that space. The spatiality of the world consists of the web of their interactions. They do not dwell in time: they interact incessantly with each other, and indeed exist only in terms of these incessant interactions. And this interaction *is* the happening of the world: it *is* the minimum elementary form of time that is neither directional nor linear. Nor does it have the curved and smooth geometry studied by Einstein. It is a reciprocal interaction in which quanta manifest themselves in the interaction, in relation to that with which they interact.

The dynamic of these interactions is probabilistic. The probabilities that something will happen – given the occurrence of something else – can in principle be calculated with the equations of the theory.

We cannot draw a complete map, a complete geometry, of everything that happens in the world, because such happenings – including the passage of time – are always triggered only by an interaction with, and with respect to, a physical system involved in the interaction. The world is like a collection of interrelated points of view. To speak of the world 'seen from outside' makes no sense, because there is no 'outside' to the world.

The elementary *quanta* of the gravitational field exist at the Planck scale. They are the elementary grains that weave the mobile fabric with which Einstein reinterpreted Newton's absolute space and time. It is these, and their interactions, which determine the extension of space and the duration of time.

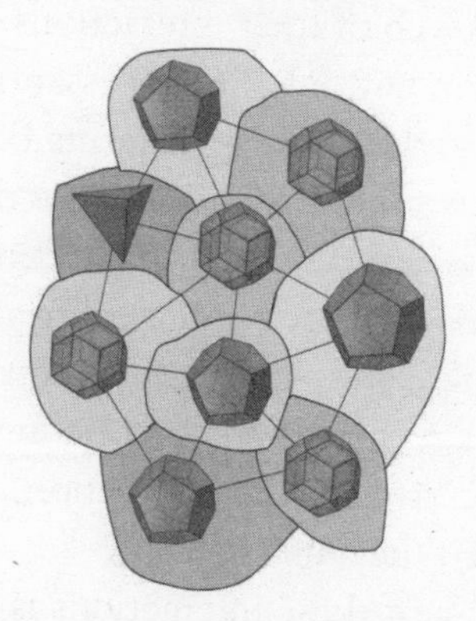

Representation of the web of elementary grains of space (or spin network).

The relations of spatial adjacency tie the grains of space into webs. We call these 'spin networks'.

The name 'spin' comes from the mathematics which describe the grains of space.[9] A ring in the spin network is called a *loop*, and these are the loops that give 'loop theory' its name.

The webs, in turn, transform into each other in discrete leaps, described in the theory as structures called 'spinfoam'.[10]

The occurrence of these leaps draws the patterns that on a large scale appear to us like the smooth structure of spacetime. On a small scale, the theory describes a 'quantum spacetime' that is fluctuating, probabilistic and discrete. At this scale, there is only the frenzied swarming of quanta that appear and vanish.

Representation of spinfoam.

This is the world with which I seek daily to come to terms. An unusual world, but not a meaningless one. In my research group in Marseille, for example, we are attempting to calculate the time needed for a black hole to explode when it passes through a quantum phase.

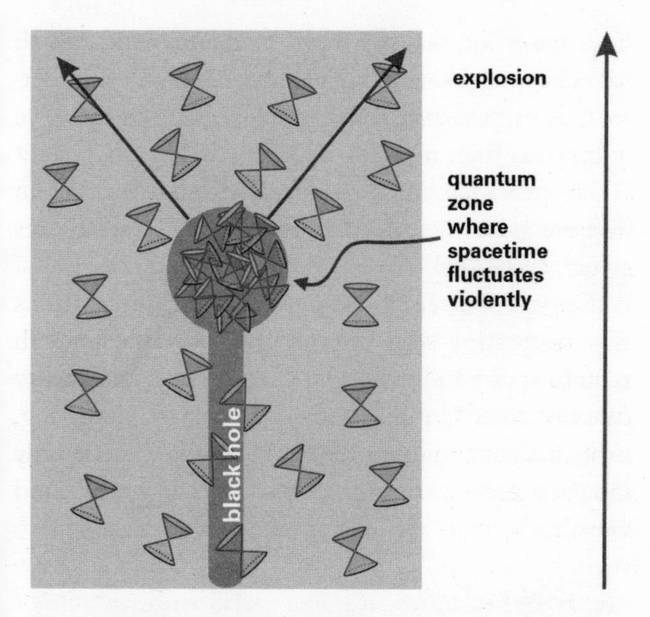

During the course of this phase, inside the black hole and within its immediate vicinity there is no longer a single and determinate spacetime. There is a quantum superposition of spin webs. Just as an electron can unfurl into a cloud of probabilities between the moment it is emitted and the moment it arrives on a screen, passing through more than one place, so the spacetime of the quantum collapse of a black hole passes through a phase in which time fluctuates violently, there is a quantum superposition of different times, and then, later, a return to a determined state after the explosion.

For this intermediate phase, where time is wholly

indeterminate, we still have equations that tell us what happens. Equations without time.

This is the world described by loop theory.

Am I certain that this is the correct description of the world? I am not, but it is today the only coherent and complete way that I know of to think about the structure of spacetime without neglecting its quantum properties. Loop quantum gravity shows that it is possible to write a coherent theory without fundamental space and time – and that it can be used to make qualitative predictions.

In a theory of this kind, time and space are no longer containers or general forms of the world. They are approximations of a quantum dynamic that in itself knows neither space nor time. There are only events and relations. It is the world without time of elementary physics.

PART THREE
The Sources of Time

9. Time is Ignorance

Do not ask
about the outcome of my days, or of yours,
Leuconoe –
it's a secret, beyond us.
And don't attempt abstruse calculations. (I, 11)

There is a time to be born and a time to die, a time to weep and a time to dance, a time to kill and a time to heal. A time to destroy and a time to build.[1] Up to this point, it has been a time to destroy time. Now it is time to rebuild the time that we experience: to look for its sources, to understand where it comes from.

If, in the elementary dynamic of the world, all the variables are equivalent, what is this thing that we humans call 'time'? What is it that my watch measures? What is it that always runs forwards, and never backwards – and why? It may not be part of the elementary grammar of the world, but what is it?

There are so many things that are not part of the elementary grammar of the world and that simply 'emerge' in some way. For example:

- A cat is not part of the elementary ingredients of the universe. It is something complex that *emerges*, and repeats itself, in various parts of our planet.
- A group of boys on a field decide to have a match. They form teams. This is how we used to do it: the two most enterprising would take turns choosing the players they wanted, having tossed a coin to see who would have first pick. At the end of this solemn procedure, there were two teams. Where were the teams before they were chosen? Nowhere. They *emerged* from the procedure.
- Where do 'high' and 'low' come from – terms that are so familiar and yet are not in the elementary equations of the world? From the Earth that is so close to us and that attracts. 'High' and 'low' *emerge* in certain circumstances in the universe, as when there is a large mass nearby.
- In the mountains, we see a valley covered by a sea of white clouds. The surface of the clouds gleams, immaculate. We start to walk towards the valley. The air becomes more humid, then less clear; the sky is no longer blue. We find ourselves in a fog. Where did the well-defined surface of the clouds go? It vanished. Its disappearance is gradual; there is no *surface* that separates the fog from the sparse air of

the heights. Was it an illusion? No, it was a view from afar. Come to think of it, it's like this with *all* surfaces. This dense marble table would look like a fog if I were shrunk to a small enough, atomic scale. *Everything* in the world becomes blurred when seen close up. Where exactly does the mountain end and where do the plains begin? Where does the savannah begin and the desert end? We cut the world into large slices. We think of it in terms of concepts that are meaningful for us, that *emerge* at a certain scale.

- We see the sky turning around us every day, but we are the ones who are turning. Is the daily spectacle of a revolving universe 'illusory'? No, it is real, but it doesn't involve the cosmos alone. It involves *our relation* with the sun and the stars. We understand it by asking ourselves how *we* move. Cosmic movement *emerges* from the relation between the cosmos and ourselves.

In these examples, something that is real – a cat, a football team, high and low, the surface of clouds, the rotation of the cosmos – *emerges* from a world that at a much simpler level has no cats, teams, up or down, no surfaces of clouds, no revolving cosmos . . . Time emerges from a world without time, in a way that has something in common with each of these examples.

The reconstruction of time begins here, in two little chapters – this one and the next – that are brief and technical. If you find them heavy going, skip them and go directly to Chapter 11. From there, step by step, we will gradually reach more human things.

Thermal Time

In the frenzy of thermal molecular mingling, all the variables that can possibly vary do so continuously.

One, however, does not vary: the total amount of energy in any isolated system. Between energy and time there is a close bond. They form one of those characteristic couples of quantities that physicists call 'conjugate', such as position and momentum, or orientation and angular momentum. The two terms of these couples are tied to each other. On the one hand, knowing what the energy of a system may be[2] – how it is linked to the other variables – is the same as knowing how time flows, because the equations of evolution in time follow from the form of its energy.[3] On the other, energy is conserved in time, hence it cannot vary, even when everything else varies. In its thermal agitation, a system[4] passes through all the configurations that have the same energy, but only these. The set of these configurations – which our blurred macroscopic vision does not distinguish – is the '(macroscopic) state of equilibrium': a placid glass of hot water.

The usual way of interpreting the relation between time and the state of equilibrium is to think that time is something absolute and objective; energy governs the time-evolution of a system; and the system in equilibrium mixes all configurations of equal energy. The conventional logic for interpreting this relation is therefore:

time → energy → macroscopic state[5]

That is: to define the macroscopic state we first need to know the energy, and to define energy we first need to know what is time. In this logic, time comes first and is independent from the rest.

But there is another way of thinking about this same relationship: by reading it in reverse. That is, to observe that a macroscopic state, which is to say a blurred vision of the world, may be interpreted as a mingling that preserves an energy, and this in its turn generates a time. That is:

macroscopic state → energy → time[6]

This observation opens up a new perspective: in an elementary physical system without any privileged variable that acts like 'time' – where, in effect, all the variables are on the same level but we can have only a blurred vision of them described by macroscopic states – a generic macroscopic state *determines* a time.

I'll repeat this point, because it is a key one: a

macroscopic state (which ignores the details) chooses a particular variable that has some of the characteristics of time.

In other words, a time becomes determined simply as an effect of blurring. Boltzmann understood that the behaviour of heat involves blurring, from the fact that inside a glass of water there is a myriad of microscopic variables that we do not see. The *number* of possible microscopic configurations for water is its entropy. But something further is also true. The blurring itself determines a particular variable: time.

In fundamental relativistic physics, where no variable plays *a priori* the role of time, we can reverse the relation between macroscopic state and evolution of time: it is not the evolution of time that determines the state, it is the state – the blurring – that determines a time.

Time that is determined in this way by a macroscopic state is called 'thermal time'. In what sense may it be said to be a time? From a microscopic point of view, there is nothing special about it – it is a variable like any other. But, from a macroscopic one, it has a crucial characteristic: among so many variables all at the same level, thermal time is the one with behaviour that most closely resembles the variable we usually call 'time', because its relations with the macroscopic states are exactly those that we know from thermodynamics.

But it is not a universal time. It is determined by a macroscopic state, that is, by a blurring, by the incompleteness of a description. In the next chapter I will discuss the origin of this blurring – but before doing so let's take another step by bringing quantum mechanics into consideration.

Quantum Time

Roger Penrose is among the most lucid of those scientists who have focused on space and time.[7] He reached the conclusion that the physics of relativity is not incompatible with our experience of the *flowing* of time but that it does not seem sufficient to account for it. He has suggested that what's missing might be what happens in a quantum interaction.[8] Alain Connes, the great French mathematician, has pointed out the deep role of quantum interaction at the root of time.

When an interaction renders the *position* of a molecule concrete, the state of the molecule is altered. The same applies for its *speed*. If what materializes *first* is the speed and *then* the position, the state of the molecule changes *in a different way* than if the order of the two events were reversed. The order matters. If I measure the position of an electron first and then its speed, its state changes differently than if I were to measure its velocity first and then its position.

This is called the 'noncommutativity' of the quantum variables, because position and speed 'do not commute', that is to say, they cannot exchange order with impunity. This noncommutativity is one of the characteristic phenomena of quantum mechanics. Noncommutativity determines an order and, consequently, a germ of temporality in the determination of two physical variables. To determine a physical variable is not an isolated act; it involves interaction. The effect of such interactions depends on their order, and this order is a primitive form of the temporal order.

Perhaps it is the very fact that the effect of these interactions depends on the order in which they take place that is at the root of the temporal order of the world. This is the fascinating idea suggested by Connes: the first germ of temporality in elementary quantum transitions lies in the fact that these interactions are naturally (partially) ordered.

Connes has provided a refined mathematical version of this idea: he has shown that a kind of temporal flow is implicitly defined by the noncommutativity of the physical variables. Due to this noncommutativity, the set of physical variables in a system defines a mathematical structure called 'noncommutative von Neumann algebra', and Connes has shown that these structures have within themselves an implicitly defined flow.[9]

Surprisingly, there is an extremely close relation between Alain Connes' flow for quantum systems and the thermal time that I have discussed above. Connes has shown that, in a quantum system, the thermal flows determined by different macroscopic states are equivalent, up to certain internal symmetries,[10] and that, together, they form precisely the Connes flow.[11]

Put more simply: the time determined by macroscopic states and the time determined by quantum noncommutativity are aspects of the same phenomenon.

And it is this thermal and quantum time, I believe,[12] that is the variable that we call 'time' in our real universe, where a time variable does not exist at the fundamental level.

The intrinsic quantum indeterminacy of things produces a blurring, like Boltzmann's blurring, which ensures – contrary to what classic physics seemed to indicate – that the unpredictability of the world is maintained even if it were possible to measure everything that is measurable.

Both the sources of blurring – quantum indeterminacy, and the fact that physical systems are composed of zillions of molecules – are at the heart of time. Temporality is profoundly linked to blurring. The blurring is due to the fact that we are ignorant of the microscopic details of the world. The time of physics is, ultimately, the expression of our ignorance of the world. Time is ignorance.

Alain Connes has co-authored with two friends a short science-fiction novel. Charlotte, the protagonist, manages to have for a moment a totality of information about the world, without blurring. She manages to 'see' the world directly, beyond time:

> I have had the unheard of good fortune of experiencing a global vision of my being – not of a particular moment, but of my existence 'as a whole'. I was able to

> compare its finite nature in space, against which no one protests, with its finite nature in time, which is instead the source of so much outrage.

And then returning to time:

> I had the impression of losing all the infinite information generated by the quantum scene, and this loss was sufficient to drag me irresistibly into the river of time.

The emotion that results from this is an emotion of time:

> This re-emergence of time seemed to me like an intrusion, a source of mental confusion, anguish, fear and alienation.[13]

Our blurred and indeterminate image of reality determines a variable, thermal time that turns out to have certain peculiar properties which begin to resemble what we call 'time': it is in the correct relation with equilibrium states.

Thermal time is tied to thermodynamics, and hence to heat, but does not yet resemble time as we experience it, because it does not distinguish between the past and the future, has no direction and lacks what we mean when we speak of its flow. We have not yet reached the time of our own experience.

The difference between the past and the future that is so important to us: where does that come from?

10. Perspective

In the impenetrable night
of his wisdom
a god closes
the strip of days
that's to come
and laughs
at our human trepidation. (III, 29)

The entire difference between past and future may be attributed solely to the fact that the entropy of the world was low in the past.[1] Why was entropy low in the past?

In this chapter I will give an account of an idea that provides a possible answer, 'if you will hear my answer to this question and its perhaps extravagant supposition'.[2] I am not sure that it is the correct answer, but it's the one with which I have become enamoured.[3] It might clarify many things.

We are the Ones Turning!

Whatever we human beings may be specifically, in detail, we are nevertheless pieces of nature, a part of the great fresco of the cosmos, a small part among many others.

Between ourselves and the rest of the world there are physical interactions. Obviously, not *all* the variables of the world interact with us, or with the segment of the world to which we belong. Only a very minute fraction of these variables does so; most of them do not react with us at all. They do not register us, and we do not register them. This is why distinct configurations of the world seem equivalent to us. The physical interaction between myself and a glass of water – two pieces of the world – is independent of the motion of the single molecules of water. In the same way, the physical interaction between myself and a distant galaxy – two pieces of the world – ignores what happens in detail out there. Therefore, our vision of the world is blurred because the physical interactions between the part of the world to which we belong and the rest are blind to many variables.

This blurring is at the heart of Boltzmann's theory.[4] From this blurring, the concepts of heat and entropy are born – and these are linked to the phenomena that characterize the flow of time. The entropy of a system depends explicitly on blurring. It depends on what I *do not* register, because it depends on the number of *indistinguishable* configurations. The *same* microscopic configuration may be of high entropy with regard to one blurring and low in relation to another.

This does not mean that blurring is a mental construct; it depends on actual, existing physical interactions.[5] Entropy is not an arbitrary quantity, or subjective. It is a *relative* one, like speed.

The speed of an object is not a property of the object alone: it is a property of the object in relation to another object. The speed of a child who is running on a moving train has a value relative to the train (a few steps per second) and a different value relative to the ground (a hundred kilometres per hour). If his mother tells the child to 'Keep still!', she does not mean that they have to throw themselves out of the window to stop *in relation to the ground*. She means that the child should stop *with regard to the train*. Speed is a property of an object with respect to another object. It is a relative quantity.

The same is true for entropy. The entropy of A with regard to B counts the number of configurations of A that the *physical* interactions between A and B do not distinguish.

Clarifying this point, which frequently causes confusion, opens up a seductive solution to the mystery of the arrow of time.

The entropy of the world does not depend only on the configuration of the world; it also depends on the way in which we are blurring the world, and this depends on what the variables of the world are that *we* interact with. That is to say, on the variables with which our part of the world interacts.

The entropy of the world in the far past appears very low to us. But this might not reflect the exact state of the world: it might regard the subset of the world's variables with which *we*, as physical systems, have interacted. It is with respect to the dramatic blurring produced by our interactions with the world, caused by the small set of macroscopic variables in terms of which we describe the world, that the entropy of the universe was low.

This, which is a *fact*, opens up the possibility that it wasn't the universe that was in a very particular configuration in the past. Perhaps instead it is us, and our interactions with the universe, that are particular. We are the ones who determine a particular macroscopic description. The initial low entropy of the universe, and hence the arrow of time, may be more down to us than to the universe itself. This is the basic idea.

Think of one of the grandest and most obvious phenomena: the diurnal rotation of the skies. It is the most immediate and magnificent characteristic of the universe around us: it turns. But is this turning really a characteristic of the universe? It is not. It took us thousands of years, but in the end we managed to understand the revolving of the heavens: we understood that it is *we* who turn, not the universe. The rotation of the heavens is a perspective effect due to our particular way of moving on Earth, rather than a mysterious property of the dynamics of the universe.

Something similar might be true for time's arrow.

The low initial entropy of the universe might be due to the particular way in which we – the physical system that we are part of – interact with it. We are attuned to a very particular subset of aspects of the universe, and it is *this* that is orientated in time.

How can a particular interaction between us and the rest of the world determine a low initial entropy?

It's simple. Take a pack of twelve cards, six red and six black. Arrange it so that the red cards are all at the front. Shuffle the pack a little and then look for the black cards that have ended up among the red ones at the front. Before shuffling, there are none; after, some. This is a basic example of the growth of entropy. At the start of the game, the number of black cards among the red in the first half of the pack is zero (the entropy is low) because it has started in a *special* configuration.

But now let's play a different game. First, shuffle the pack in a random way, then look at the first six cards and commit them to memory. Shuffle a little and look to see which other cards have ended up among the first six. At the start, there were none, then their number grew, as it did in the previous example, together with the entropy. But there is a crucial difference between this example and the previous one: at the beginning of this one, the cards were in a *random* configuration. It was *you* who declared them to be particular, by taking note of which cards were in the front half of the pack at the beginning of the game.

The same may be true for the entropy of the

universe: perhaps it was in no particular configuration. Perhaps we are the ones who belong to a particular physical system with respect to which its state can be particular.

But why should there be such a physical system, in relation to which the initial configuration of the universe turns out to be special? Because in the vastness of the universe there are innumerable physical systems, and they interact with each other in ways that are even more numerous. Among these, through the endless game of probabilities and huge numbers, there will surely be some that interact with the rest of the universe *precisely* with those variables that found themselves having a particular value in the past.

It is hardly surprising that there are 'special' subsets in a universe as vast as ours. It is not surprising that *someone* wins the lottery: someone wins it every week. It is unnatural to assume that the entire universe has been in an incredibly 'special' configuration in the past, but there is nothing unnatural in imagining that the universe has parts that are 'special'.

If a subset of the universe is special in this sense, then for this subset the entropy of the universe is low in the past, the second law of thermodynamics obtains; memories exist, traces are left – and there can be evolution, life and thought.

In other words, if in the universe there is something like this – and it seems natural to me that there could be – then we belong to that something. Here, 'we' refers to that collection of physical variables to which we commonly have access and by means of which we describe the universe. Perhaps, therefore, the flow of time is not a characteristic of the universe: like the rotation of the heavens, it is due to the particular perspective that we have from our corner of it.

But why should *we* belong to one of *these* special systems? For the same reason that apples grow in northern Europe, where people drink cider, and grapes grow in the south, where people drink wine; or that I was born where people happen to speak my native language; or that the sun which warms us is at the right distance from us – not too close and not too far away. In all these cases, the 'strange' coincidence arises from confusing the causal relations: it isn't that apples grow where people drink cider, it is that people drink cider where apples grow. Put this way, there is no longer anything strange about it.

Similarly, in the boundless variety of the universe, it may happen that there are physical systems that interact with the rest of the world through those particular variables that define an initial low entropy. With regard to *these* systems, entropy is constantly increasing. There, and not elsewhere, there are the typical phenomena associated with the flowing of time: life is possible, together with evolution, thought and our awareness of time passing. There, the apples grow that produce our cider: time. That sweet juice which contains all the ambrosia and all the gall of life.

Indexicality

When we do science, we want to describe the world in the most objective way possible. We try to eliminate distortions and optical illusions deriving from our point of view. Science aspires to objectivity, to a shared point of view about which it is possible to be in agreement.

This is admirable, but we need to be wary about what we lose by ignoring the point of view from which we do the observing. In its anxious pursuit of objectivity, science must not forget that our experience of the world comes from within. Every glance that we cast towards the world is made from a particular perspective.

Taking this fact into account helps to clarify many things. It clarifies, for instance, the relation between what a geographical map tells us and what we actually see. In order to compare the map with what we see, it is necessary to add a crucial piece of information: we must identify on the map our exact location. The map does not know where we are, at least when it is not fixed in the place that it represents – like those maps in mountain villages showing routes that can be walked, with a red dot next to which is written: 'You are Here.'

A strange phrase: how can a map know where we are? We might be looking at it from afar, through binoculars. Instead, it should say, 'I, a map, am here,' with an arrow next to the red dot. But there is also something curious about a text which refers to itself. What is it?

It is what philosophers call 'indexicality': the characteristic of certain words which have a different meaning every time they are used, a meaning determined by where, how, when and by whom they are being spoken. Words such as 'here', 'now', 'I', 'this', 'tonight' all assume a different meaning depending on who utters them and the circumstances in which they are uttered. 'My name is Carlo Rovelli' is true if I say it, but untrue if someone else not also called Carlo Rovelli uses the same phrase. 'Now it is 12 September 2016' is a phrase that's true at the moment that I am writing this but will be false just a few hours later. These indexical phrases make explicit reference to the fact that point of view exists, that a point of view is an ingredient in every description of the observable world that we make.

If we give a description of the world that ignores point of view, that is solely 'from the outside' – of space, of time, of a subject – we may be able to say many things but we lose certain crucial aspects of the world. Because the world that we have been given is the world seen from within it, not from without.

Many things that we see in the world can be understood only if we take into account the role played by point of view. They remain unintelligible if we fail to do so. In every experience, we are situated within the world: within a mind, a brain, a position in space, a moment in time. Our being situated in the world is essential to understanding our experience of time. We must not, in short, confuse the temporal structures that belong to the world as 'seen from the outside' with the aspects of the world that we observe and which depend on our being part of it, on our being situated within it.[6]

In order to use a geographical map, it is not enough to look at it from the outside: we must know where we are situated in relation to what it represents. In order to understand our experience of space, it is not enough to think of Newtonian space. We must remember that we see this space from inside it, that we are localized. In order to understand time, it is not enough to think of it from outside: it is necessary to understand that we, in every moment of our experience, are situated *within* time.

We observe the universe from within it, interacting with a minuscule portion of the innumerable variables of the cosmos. What we see is a blurred image. This blurring suggests that the dynamic of the universe with which we interact is governed by entropy, which measures the amount of blurring. It measures something that relates to us more than to the cosmos.

We are getting up dangerously close to ourselves. We can almost hear Tiresias, in *Oedipus*, saying: 'Stop! Or you will find yourself' ... Or Hildegard of

Bingen, who in the twelfth century seeks the absolute and ends up by putting 'universal man' at the centre of the cosmos.

But, before getting to this 'us', another chapter is required, to show how the growth in entropy – perhaps only an effect of perspective – might give rise to the entire, vast phenomenon of time.

Let me summarize the hard ground covered in the last two chapters, in the hope that I have not already

Universal man at the centre of the cosmos, in Liber Divinorum Operum *(1164–70) by Hildegard of Bingen.*

lost all my readers. At the fundamental level, the world is a collection of events *not* ordered in time. These events manifest relations between physical variables that are, *a priori*, on the same level. Each part of the world interacts with a small part of all the variables, the value of which determines 'the state of the world with regard to that particular subsystem'.

A small system S does not distinguish the details of the rest of the universe, because it interacts only with a few among the variables of the rest of the universe. The entropy of the universe *with respect to* S counts the (micro) states of the universe undistinguishable *by* S. The universe appears in a high-entropy configuration *with respect to* S, because (by definition) there are more microstates in high-entropy configurations, and therefore it is more likely to happen to be in one of these microstates.

As explained above, there is a *flow* associated with high-entropy configurations and the parameter of this flow is *thermal time*. For a generic small system S, entropy remains generally high along the entire flow of thermal time, perhaps just fluctuating up and down, because, after all, we are dealing here with probabilities, not fixed rules.

But among the innumerable small systems S that exist in this extraordinarily vast universe where we happen to live, there will be a few special ones for which the fluctuations of the entropy happen to be such that *at one of the two ends* of the flow of thermal time entropy happens to be low. For *these* systems S, the fluctuation is not symmetrical: entropy increases. This growth is what we experience as the flowing of time. What is special is not the state of the early universe: it is the small system S to which we belong.

I'm not sure if we are dealing with a plausible story, but I do not know of any better ones. The alternative is to accept as a given of observation the fact that entropy was low at the beginning of the universe, and to leave it at that.[7]

It is the law announced by Clausius, $\Delta S \geq 0$, and deciphered by Boltzmann that is guiding us: entropy never decreases. Having lost sight of it, in search of the general laws of the world, we rediscover it as a possible perspective effect for particular subsystems. Let's begin again from there.

11. What Emerges from a Particularity

Why do the tall pine
and the pale poplar
intertwine their branches
to provide such sweet shade for us?
Why does the fleeting water
invent bright spirals
in the turbulent stream? (II, 9)

It is Entropy, not Energy, that Drives the World

At school I was told that it is energy that makes the world go round. We need to get energy, for example from petroleum, from the sun or from nuclear sources. Energy makes our engines run, helps plants to grow and causes us to wake up every morning full of vitality.

But there is something that does not add up. Energy – as I was also told at school – is conserved. It is neither created nor destroyed. If it is conserved, why do we have to constantly resupply it? Why can't we just keep using the same energy?

The truth is that there is plenty of energy and it is not consumed. It's not energy that the world needs in order to keep going. What it needs is low entropy.

Energy (be it mechanical, chemical, electrical or potential) transforms itself into thermal energy, that is to say, into heat: it goes into cold things, and there is no free way of getting it back from there to reuse it to make a plant grow or to power a motor. In this process, the energy remains the same but the entropy increases, and it is *this* which cannot be turned back. The second law of thermodynamics demands it.

What makes the world go round are not sources of energy but sources of low entropy. Without low entropy, energy would dilute into uniform heat and the world would go to sleep in a state of thermal equilibrium – there would no longer be any distinction between past and future, and nothing would happen.

Near to the Earth we have a rich source of low entropy: the sun. The sun sends us hot photons. Then the Earth radiates heat towards the black sky, emitting colder photons. The energy that enters is more or less equal to the energy that exits; consequently, we do not generally gain energy in the exchange. (Gaining energy in the exchange is disastrous for us: it is global warming.) But for every hot photon that arrives, the Earth emits ten cold ones, since a hot photon from the sun has the same energy as ten cold photons emitted by the Earth. The hot photon has *less entropy* than the ten cold photons, because the number of configurations of a single (hot) photon is lower than the number of configurations of ten (cold) photons. Therefore, the sun is a continual rich source of low entropy for us. We have at our disposal an abundance of low entropy, and it is this that allows plants and animals to grow, enables us to build motors and cities – and to think and to write books such as this one.

Where does the low entropy of the sun come from? From the fact that, in turn, the sun is born out of an entropic configuration that was even lower: the primordial cloud from which the solar system was formed had even lower entropy. And so on, back into the past, until we reach the extremely low initial entropy of the universe.

It is the growth of this entropy that powers the great story of the cosmos.

But the increase in the entropy of the universe is not rapid, like the sudden expansion of gas in a box: it is gradual, it takes time. Even with a gigantic ladle it takes time to stir something as big as the universe. Above all, there are obstacles and closed doors to its growth – passages where it occurs only with great difficulty.

A pile of wood, for example, lasts a long time if left alone. It is not in a state of maximum entropy, because the elements of which it is made, such as carbon and hydrogen, are combined in a very particular manner ('ordered') to give form to the wood. Entropy grows if these particular combinations are broken down.

This is what happens when wood burns: its elements disengage from the particular structures that form wood and entropy increases sharply (fire being, in fact, a markedly irreversible process). But the wood does not start to burn on its own. It remains for a long time in a state of low entropy, until something opens a door that allows it to pass to a state of higher entropy. A pile of wood is in an unstable state, like a pack of cards, but until something comes along to make it do so, it does not collapse. This something might, for instance, be a match to light a flame. The flame is a process that opens a channel through which the wood can pass into a state of higher entropy.

There are situations which impede and hence slow down the increase of entropy throughout the universe. In the past, for instance, the universe was largely an immense expanse of hydrogen. Hydrogen can fuse into helium, and helium has a higher entropy than hydrogen. But for this to happen it is necessary for a channel to be opened: a star must ignite for hydrogen to begin to burn there into helium. What causes stars to ignite? Another process that increases entropy: the contraction due to gravity of one of the large clouds of hydrogen that sail throughout the galaxy. A contracted cloud of hydrogen has higher entropy than a dispersed one.[1] But the clouds of hydrogen are so vast that they take millions of years to contract. Only after they have become concentrated do they manage to heat up to the point that triggers the process of nuclear fusion.

The ignition of nuclear fusion opens the door that allows the further increase in entropy: hydrogen burning into helium.

The entire history of the universe consists of this halting and leaping cosmic growth of entropy. It is neither rapid nor uniform, because things remain trapped in basins of low entropy (the pile of wood, the cloud of hydrogen . . .) until something opens a door on to a process that finally allows entropy to increase. The growth of entropy itself happens to open new doors through which entropy can increase further. A dam in the mountains, for instance, retains water until it is gradually worn down over time and the freed water escapes downhill once again, causing entropy to grow. Over the course of this irregular trajectory, large or small portions of the universe remain isolated in relatively stable situations for periods that can be very prolonged.

Living beings are made up of similarly intertwined processes. Photosynthesis deposits low entropy from the sun into plants. Animals feed on low entropy by eating. (If all we needed was energy rather than entropy, we would head for the heat of the Sahara rather than towards our next meal.) Inside every living cell, the complex web of chemical processes is a structure that opens and closes gates through which low entropy can increase. Molecules function as the catalysts that allow the processes to intertwine; or, conversely, they put a brake on them. The increase of entropy in each individual process is what makes the whole thing work. Life is this network of processes for increasing entropy – processes which act as catalysts to each other.[2] It isn't true, as is sometimes stated, that life generates structures that are particularly ordered, or that locally diminish entropy: it is simply a process that degrades and consumes the low entropy of food; it is a self-structured disordering, no more and no less than in the rest of the universe.

Even the most banal phenomena are governed by the second law of thermodynamics. A stone falls to the ground. Why? One often reads that it's because the stone places itself 'in a state of lower energy' that it ends up lower down. But why does the stone put itself into a state of lower energy? Why should it lose energy if energy is conserved? The answer is that when the stone hits the Earth, it warms it: its mechanical energy is transformed into heat. And there is no way back from there. If the second law of thermodynamics did not exist, if heat did not exist, if there existed no microscopic swarming, the stone would rebound perpetually; it would never land and be still.

It is entropy, not energy, that keeps stones on the ground and the world turning.

The entire coming into being of the cosmos is a gradual process of disordering, like the pack of cards that begins in order and then becomes disordered through shuffling. There are no immense hands that shuffle the universe. It does this mixing by itself, in the interactions between its parts that open and close during the course of the mixing, step by step. Vast regions remain trapped in configurations that remain ordered, until here and there new channels are opened through which disorder spreads.[3]

What causes events to happen in the world, what writes its history, is the irresistible mixing of all things, going from the few ordered configurations to the countless disordered ones. The entire universe is like a mountain that collapses in slow motion. Like a structure that very gradually crumbles.

From the most minute events to the more complex ones, it is this dance of ever-increasing entropy, nourished by the initial low entropy of the universe, that is the real dance of Shiva, the destroyer.

Traces and Causes

The fact that entropy has been low in the past leads to an important fact that is ubiquitous and crucial for the difference between past and future: the past leaves traces of itself in the present.

Traces are everywhere. The craters of the moon testify to impacts in the past. Fossils show the forms of living creatures from long ago. Telescopes show how far off galaxies were in the past. Books contain our history; our brains swarm with memories.

Traces of the past exist, and not traces of the future,

only because entropy was low in the past. There can be no other reason, since the only source of the difference between past and future is the low entropy of the past.

In order to leave a trace, it is necessary for something to become arrested, to stop moving, and this can happen only in an irreversible process – that is to say, by degrading energy into heat. In this way, computers heat up, the brain heats up, the meteors that fall into the moon heat it; even the goose quill of a medieval scribe in a Benedictine abbey heats a little the page on which he writes. In a world without heat, everything would rebound elastically, leaving no trace.[4]

It is the presence of abundant traces of the past that produces the familiar sensation that the past is determined. The absence of any analogous traces of the future produces the sensation that the future is open. The existence of traces serves to make it possible for our brain to create extensive maps of past events. There is nothing analogous to this for future ones. This fact is at the origin of our sensation of being able to act freely in the world: choosing between different futures, even though we are unable to act upon the past.

The vast mechanisms of the brain about which we have no direct awareness ('I know not why I am so sad,' mumbles Antonio at the beginning of *The Merchant of Venice*) have been designed during the course of evolution in order to make calculations about possible futures. This is what we call 'deciding'. Since they elaborate possible alternative futures that would follow if the present were exactly as it is except for some detail, we are naturally inclined to think in terms of 'causes' that precede 'effects': the cause of a future event is a past event such that the future event would not follow in a world that was exactly the same except for this cause.[5]

In our experience, the notion of cause is thus asymmetrical in time: cause precedes effect. When we recognize in particular that two events 'have the same cause', we find this common cause[6] in the past, not in the future. If two waves of a tsunami arrive together at two neighbouring islands, we think that there has been an event *in the past* that has caused both. We do not look for it in the future. But this does not happen because there is a magical force of 'causality' going from the past to the future. It happens because the improbability of a correlation between two events requires something improbable, and it is *only* the low entropy of the past that provides such improbability. What else could? In other words, the existence of common causes in the past is nothing but a manifestation of low entropy in the past. In a state of thermal equilibrium, or in a purely mechanical system, there isn't a direction to time identified by causality.

The laws of elementary physics do not speak of 'causes' but only of regularities, and these are symmetrical with regard to past and future. Bertrand Russell noted this in a famous article, writing emphatically that, 'The law of causality . . . is a relic of a bygone age, surviving, like the monarchy, only because it is erroneously supposed to do no harm.'[7] He exaggerates, of course, because the fact that there are no 'causes' *at an elementary level* is not a sufficient reason to render obsolete the very notion of cause.[8] At an elementary level there are no cats either, but we do not for this reason cease to bother with cats. The low entropy of the past renders the notion of cause an effective one.

But memory, causes and effects, flow, the determined nature of the past and the indeterminacy of the future are nothing but names that we give to the consequences of a statistical fact: the improbability of a past state of the universe.

Causes, memory, traces, the history itself of the becoming of the world that unfolds not only across centuries and millennia of human history but in the billions of years of the great cosmic narrative – all this stems simply from the fact that the configuration of things was 'particular' a few billion years ago.[9]

And 'particular' is a relative term: it is particular in relation to a perspective. It is a blurring. It is determined by the interactions that a physical system has with the rest of the world. Hence causality, memory, traces, the history of the happening of the world itself can only be an effect of perspective: like the turning of the heavens; an effect of our peculiar point of view in the world . . . Inexorably, then, the study of time does nothing but return us to ourselves.

12. The Scent of the Madeleine

Happy
and master of himself
is the man who
for every day of his life can say:
'Today I have lived;
tomorrow if God extends for us
a horizon of dark clouds
or designs a morning
of limpid light,
he will not change our poor past,
he will do nothing without the memory
of events that the fleeting hour
will have assigned to us.' (III, 29)

Let us turn to ourselves, then, and to the role we play in relation to the nature of time. Above all else, what *are* we as human beings? Entities? But the world is not made up of entities, it is made from events that combine with each other . . . So what, then, am 'I'?

In the *Milinda Pānha*, a Buddhist text written in Pali in the first century of our era, Nāgasena replies to the questions of King Milinda, denying his existence as an entity:[1]

> King Milinda says to the sage Nāgasena: What is your name, Master? The teacher replies: I am called Nāgasena, o great king; Nāgasena is nothing but a name, a designation, an expression, a simple word: there is no person here.

The king is astonished by such an extreme-sounding assertion:

> If no person exists, who is it then who has clothing and sustenance? Who lives according to the virtues? Who kills, who steals, who has pleasures, who lies? If there is no longer an actor, neither is there good or evil any longer . . .

And he argues that the subject must be an autonomous being that is not reducible to its component parts:

> Is it the hairs that are Nāgasena, Master? Is it the nails or the teeth or the flesh or the bones? Is it the name? Is it the sensations, the perceptions, the consciousness? Is it none of these things? . . .

The sage replies that 'Nāgasena' is effectively none of these things, and the king seems to have won the discussion: if Nāgasena is none of these, then he must be something else – and this something else will be the person Nāgasena who therefore exists.

But the sage turns his own argument against him, asking what a chariot consists of:

> Are the wheels the chariot? Is the axle? Is the chassis the chariot? Is the chariot the sum of its parts?

The king replies cautiously that certainly 'chariot' refers only to the relationship between the ensemble of wheels, axle and chassis, to their working together and in relation to us – and that there does not exist an entity 'chariot' beyond these relations and events. Nāgasena triumphs: in the same way as 'chariot', the name 'Nāgasena' designates nothing more than a collection of relations and events.

We are processes, events, composite and limited in space and time. But, if we are not an individual entity, what is it that founds our identity and its unity? What makes it so – that I am Carlo – and that my hair and my nails and my feet are considered part of me, as well as my anger and my dreams, and that I consider myself to be the same Carlo as yesterday, the same as tomorrow; the one who thinks, suffers and perceives?

There are different ingredients that combine to produce our identity. Three of these are important for the argument of this book:

1.

The first is that every one of us identifies with a *point of view* in the world. The world is reflected in each one of us through a rich spectrum of correlations essential for our survival.[2] Each of us is a complex process that reflects the world and elaborates the information we receive in a way that is strictly integrated.[3]

2.

The second ingredient on which our identity is based is the same as for the chariot. In the process of reflecting the world, we organize it into entities: we conceive of the world by grouping and segmenting it as best we can in a continuous process that is more or less uniform and stable, the better to interact with it. We group together into a single entity the rocks that we call Mont Blanc, and we think of it as a unified thing. We draw lines over the world, dividing it into sections; we establish boundaries, we approximate the world by breaking it down into pieces. It is the structure of our nervous system that works in this way. It receives sensory stimuli, elaborates information continuously, generating behaviour. It does so through networks of neurons which form flexible dynamic systems that continuously modify themselves, seeking to predict[4] – as far as possible – the flow of information intake. In order to do this, the networks of neurons evolve by associating more or less stable fixed points of their dynamic to recurring patterns that they find in the incoming information, or – indirectly – in the procedures of elaboration themselves. This is what seems to emerge from the very lively current research on the brain.[5] If this is so, then 'things', like 'concepts', are fixed points in the neuronal dynamic, induced by recurring structures of the sensorial input and of the successive elaborations. They mirror a combination of aspects of the world that depends on recurrent structures of the world and on their relevance in their interactions with us. This is what a chariot consists of. Hume would have been pleased to know about these developments in our understanding of the brain.

In particular, we group into a unified image the collection of processes that constitutes those living organisms that are *other* human beings, because our life is social and we therefore interact a great deal with them. They are knots of cause and effect that are deeply relevant for us. We have shaped an idea of a 'human being' by interacting with others like ourselves.

I believe that our notion of self stems from this, not from introspection. When we think of ourselves as persons, I believe we are applying to ourselves the mental circuits that we have developed to engage with our companions.

The first image that I have of myself as a child is the child that my mother sees. We are for ourselves in large measure what we see and have seen of ourselves reflected back to us by our friends, our loves and our enemies.

I have never been convinced by the idea, attributed to Descartes, that the primary aspect of our experience is awareness of thinking, and therefore of existing. (Even the attribution of the idea to Descartes seems wrong to me: *Cogito ergo sum* is not the first step in the Cartesian reconstruction, it is the second. The first is *Dubito ergo cogito*. The starting point of the reconstruction is not a hypothetical *a priori* that is immediate to the experience of existing as a subject. It's a rationalistic *a posteriori* reflection on the first stage of the process in which Descartes had articulated a state of doubt: logic dictates that, if someone doubts something, they must have thought about it. And that, if they can think, then they must exist. It is substantially a consideration made in the third person, not in the first, however private the process. The starting point for Descartes is the methodical doubt experienced by a refined intellectual, not the basic experience of a subject.)

The experience of thinking of oneself as a subject is not a primary experience: it is a complex cultural deduction, made on the basis of many other thoughts. My primary experience – if we grant that this means anything – is to see the world around me, not myself. I believe that we each have a concept of 'my self' only because at a certain point we learn to project on to ourselves the idea of being human as an additional feature that evolution has led us to develop during the course of millennia in order to engage with other members of our group: we are the reflection of the idea of ourselves that we receive back from our kind.

3.

But there is a third ingredient in the foundation of our identity, and it is probably the essential one – it is the reason this delicate discussion is taking place in a book about time: memory. We are not a collection of independent processes in successive moments. Every moment of our existence is linked by a peculiar triple thread to our past – the most recent and the most distant – by memory. Our present swarms with traces of our past. We are *histories* of ourselves. Narratives. I am not this momentary mass of flesh reclined on the sofa typing the letter 'a' on my laptop; I am my thoughts full of the traces of the phrases that I am writing; I am my mother's caresses, and the serene kindness with which my father calmly guided me; I am my adolescent travels; I am what my reading has deposited in layers in my mind; I am my loves, my moments of despair, my friendships, what I've written, what I've heard; the faces engraved on my memory. I am, above all, the one who a minute ago made a cup of tea for himself. The one who a moment ago typed the word 'memory' into his computer. The one who just composed the sentence that I am now completing. If all this disappeared, would I still exist? I am this long, ongoing novel. My life consists of it.

It is memory that solders together the processes, scattered across time, of which we are made. In this sense we exist in time. It is for this reason that I am the same person today as I was yesterday. To understand ourselves means to reflect on time. But to understand time we need to reflect on ourselves.

A recent book devoted to research on the functioning of the brain is entitled *Your Brain is a Time Machine.*[6] It discusses the many ways in which the brain interacts with the passage of time and establishes bridges between past, present and future. To a large extent, the brain is a mechanism for collecting memories of the past in order to use them continually to predict the future. This happens across a wide spectrum of time scales, from the very short to the very long. If someone throws something at us to catch, our hand moves skilfully to the place where the object will be in a few instants: the brain, using past impressions, has very rapidly calculated the future position of the object that is flying towards us. Further along the scale, we plant seed so that corn will grow. Or invest in scientific research so that tomorrow it might result in knowledge and new technology. The possibility of predicting something in the future obviously improves our chances of survival and, consequently, evolution has selected the neural structures that allow it. We are the result of this selection. This being between past and future events is central to our mental structure. This, for us, is the 'flow' of time.

There are elementary structures in the wiring of our nervous system that immediately register movement: an object that appears in one place and then immediately afterwards in another does not generate two distinct signals that travel separately towards the brain but a single signal correlated with the fact that we are looking at something that is moving. In other words, what we perceive is not the present, which in any case makes no sense for a system that functions on a scale of finite time, but rather something that happens and extends in time. It is in our brains that an extension in time becomes condensed into a perception of duration.

This intuition is an ancient one. St Augustine's ruminations on it have remained famous.

In Book XI of the *Confessions*, Augustine asks himself about the nature of time and, despite being interrupted by exclamations in the style of an evangelical preacher that I find quite tiresome, he presents a lucid analysis of our capacity for perceiving time. He observes that we are always in the present, because the past is past and therefore does not exist, and the future has yet to arrive, so it does not exist either. And he asks himself how we can be aware of duration – or even be capable of evaluating it – if we are always only in a present which is, by definition, instantaneous.

How can we come to know so clearly about the past, about time, if we are always in the present? Here and now, there is no past and no future. Where are they? Augustine concludes that they are within us:

> It is within my mind, then, that I measure time. I must not allow my mind to insist that time is something objective. When I measure time, I am measuring something in the present of my mind. Either this is time, or I have no idea what time is.

The idea is much more convincing than it seems on first reading. We can say that we measure duration with a clock. But to do so requires us to read it at two different moments: this is not possible, because we are always in one moment, never in two. In the present, we see only the present; we can see things that we interpret as *traces* of the past, but there is a categorical difference between seeing traces of the past and perceiving the flow of time – and Augustine realizes that the root of this difference, the awareness of the passing of time, is internal. It is integral to the mind. It is the traces left in the brain by the past.

Augustine's exposition of the idea is quite beautiful. It is based on our experience of music. When we listen to a hymn, the meaning of a sound is given by the ones that come before and after it. Music can occur only in time, but if we are always in the present moment, how is it possible to hear it? It is possible,

Augustine observes, because our consciousness is based on memory and on anticipation. A hymn, a song, is in some way present in our minds in a unified form, held together by something – by that which we take time to be. And hence this is what time is: it is entirely in the present, in our minds, as memory and as anticipation.

The idea that time might exist only in the mind certainly did not become dominant in Christian thought. In fact, it is one of the propositions explicitly condemned as heretical by the Bishop of Paris Étienne Tempier in 1277. In his list of beliefs to be condemned, the following can be found:

> *Quod evum et tempus nichil sunt in re, sed solum in apprehensione.*[7]

In other words: '[It is heretical to maintain that] the age and time do not exist in reality but only in the mind.' Perhaps my book is sliding towards heresy . . . But, given that Augustine continues to be regarded as a saint, I don't suppose that I should be too worried about it. Christianity, after all, is quite flexible . . .

It may seem easy to refute Augustine by arguing that the traces of the past which he finds within himself may be there only because they reflect a real structure of the external world. In the fourteenth century, for instance, William of Ockham maintained in his *Philosophia Naturalis* that man observes both the sky's movements and the ones within himself and therefore perceives time through his coexistence with the world. Centuries later, Husserl insists – rightly – on the distinction between physical time and 'internal consciousness of time': for a sound naturalist wishing to avoid drowning in the useless vortices of idealism, the former (the physical world) comes first, while the latter (consciousness) – independently of how well we understand it – is determined by the first. It is an entirely reasonable objection, just so long as physics reassures us that the external flow of time is real, universal and in keeping with our intuitions. But, if physics shows us instead that such time is *not* an elementary part of reality, can we continue to overlook Augustine's observation and treat it as irrelevant to the true nature of time?

Inquiry into perception of the *internal* rather than *external* nature of time reoccurs frequently in Western philosophy. Kant discusses the nature of space and time in his *Critique of Pure Reason*, and interprets both space and time as *a priori* forms of knowledge, that is to say, things that relate not just to the objective world but also to the way in which a subject apprehends it. But he also observes that, whereas space is shaped by our *external* sense, that is to say, by our way of ordering things that we see in the world *outside* of us, time is shaped by our *internal* sense, that is to say, by our way of ordering internal states *within* ourselves. Once again: the basis of the temporal structure of the world is to be sought in something that closely relates to our way of thinking and perceiving, to our consciousness. This remains true without having to get tangled up in Kantian transcendentalism.

Husserl reprises Augustine when he describes the shaping of experience in terms of 'retention' – using, like him, the metaphor of listening to a melody[8] (the world, in the meantime, has become bourgeois, with melodies replacing hymns): in the moment that we hear a note the previous note is 'retained', then that one also becomes part of the retention – and so on, running them together in such a way that the present contains a continuous trace of the past, becoming gradually more blurred.[9] It is by way of this process of retention, according to Husserl, that phenomena 'constitute time'. The following diagram is Husserl's: the horizontal axis from A to E represents time passing, the vertical axis from E to A′ represents the 'retention' of moment A, where the progressive subsidence leads from A to A′. Phenomena constitute time because, at any moment E, P′ and A′ exist. The interesting point here is that the source of the phenomenology of time is not identified by Husserl in the hypothetical objective succession of phenomena (the horizontal line) but rather in memory (similar to anticipation, called 'protention' by Husserl), that is to say, by the vertical line in the diagram. My point is that this continues to be valid (in a natural philosophy) even in a physical world where there is no physical time globally organized in a linear way but only traces generated by varying entropy.

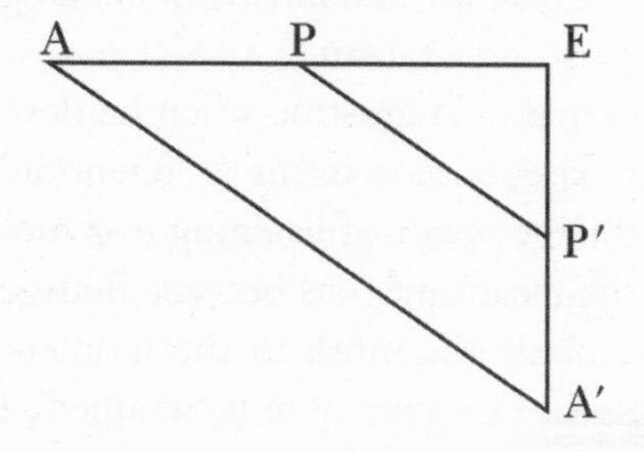

In the wake of Husserl, Martin Heidegger writes – as far as my love of the clarity and transparency of Galileo's writing allows me to decipher the deliberate obscurity of Heidegger's language – that 'time temporalizes itself only to the extent that it is human'.[10] For him also, time is the time of mankind, the time for doing, for that with which mankind is engaged. Even if, afterwards, since he is interested in what being is for man (for 'the entity that poses the problem of existence'),[11] Heidegger ends up by identifying the internal consciousness of time as the horizon of being itself.

These intuitions of the degree to which time is inherent to subjectivity remain significant also to any sound naturalism that sees the subject as part of nature and is not afraid to speak about 'reality' and to study it – while at the same time acknowledging that our understanding and our intuition are radically filtered by the way in which that limited instrument – our brain – works. This brain is part of that reality which consequently depends on the interaction between an external world and the structures with which the mind operates.

But the mind is the working of our brain. What (little) we are beginning to understand of this functioning is that our entire brain operates on the basis of a collection of *traces* of the past left in the synapses that connect neurons. Synapses are continually formed in their thousands and then erased – especially during sleep, leaving behind a blurry reflection of that which has acted on our nervous system in the past. A blurred image, no doubt – think of how many millions of details our eyes see every moment which do *not* stay in our memory – but one which contains worlds.

Boundless worlds.

They are those worlds that the young Marcel rediscovers, bewildered, every morning, in the first pages of *Remembrance of Things Past*, in the vertigo of the moment when consciousness emerges like a bubble from unfathomable depths.[12] That world of which vast territories are then revealed to him when the taste of the madeleine brings back the flavour of Combray. A vast world, a map of which Proust slowly unfolds during the course of the three thousand pages of his great novel. A novel, it should be noted, that is not a narrative of events in the world but an account of what's inside the memory of a single person. From the fragrance of the madeleine at the beginning, to the last word – 'time' – of its final part, *Time Regained*, the book is nothing but a disordered, detailed meandering among the synapses of Marcel's brain.

Proust finds a limitless space and an incredible throng of details, fragrances, considerations, sensations, reflections, re-elaborations, colours, objects, names, looks, emotions . . . all within the folds of the brain between the ears of Marcel. This is the flow of time familiar from our experience: it is inside there that it nestles, inside of us, in the utterly crucial presence of traces of the past in our neurons.

Proust could not be more explicit on this matter, writing in the first book: 'Reality is formed only by memory.'[13] And memory, in its turn, is a collection of traces, an indirect product of the disordering of the world, of that small equation written earlier, $\Delta S \geq 0$, the one that tells us the state of the world was in a 'particular' configuration in the past and therefore has left (and leaves) traces. 'Particular', that is, perhaps only in relation to rare subsystems – ourselves included.

We are stories, contained within the twenty complicated centimetres behind our eyes, lines drawn by traces left by the (re)mingling together of things in the world, and orientated towards predicting events in the future, towards the direction of increasing entropy, in a rather particular corner of this immense, chaotic universe.

This space – memory – combined with our continuous process of anticipation, is the source of our sensing time as time, and ourselves as ourselves.[14] Think about it: our introspection is easily capable of imagining itself without there being space or matter, but can it imagine itself not existing in time?[15]

It is with respect to that physical system to which we belong, due to the peculiar way in which it interacts with the rest of the world, thanks to the fact that it allows traces and because we, as physical entities, consist of memory and anticipation, that the perspective of time opens up for us, like our small, lit clearing.[16] Time opens up our limited access to the world.[17] Time, then, is the form in which we beings whose brains are made up essentially of memory and foresight interact with the world: it is the source of our identity.[18]

And of our suffering as well.

Buddha summed this up in a few maxims that millions of human beings have adopted as the foundations of their lives: birth is suffering, decline is suffering, illness is suffering, death is suffering, union with that which we hate is suffering, separation from that which we love is suffering, failure to obtain what we desire is suffering.[19] It's suffering because we must lose what we have and are attached to. Because everything that begins must end. What causes us to suffer is not in the past or the future: it is there, now, in our memory, in our expectations. We long for timelessness, we endure the passing of time: we suffer time. Time is suffering.

Such is time, and because of this we are fascinated and troubled by it in equal measure – and perhaps because of this, too, dear reader, my brother, my sister, you are holding this book in your hands. Because it is nothing but a fleeting structure of the world, an ephemeral fluctuation in the happening of the world, that which is capable of giving rise to what we are: beings made of time. That to which we owe our being, giving us the precious gift of our very existence, allowing us to create the fleeting illusion of permanence that is the origin of all our suffering.

The music of Strauss and the words of Hofmannsthal sing of this with devastating delicacy:[20]

> I remember a little girl . . .
> But how can that be . . .
> Once I was that little Resi,
> and then one day I became an old woman?
> . . . If God wills it so, why allow me to see it?
> Why doesn't he hide it from me?
> Everything is a mystery, such a deep mystery . . .
> I feel the fragility of things in time.
> From the bottom of my heart, I feel we
> should cling to nothing.

Everything slips through our fingers.
All that we seek to hold on to dissolves.
Everything vanishes, like mist and dreams . . .
Time is a strange thing.
When we don't need it, it is nothing.
Then, suddenly, there is nothing else.
It is everywhere around us. Also within us.
It seeps into our faces.
It seeps into the mirror, runs through my temples . . .
Between you and I it runs silently, like an hourglass.
Oh, Quin Quin.
Sometimes I feel it flowing inexorably.
Sometimes I get up in the middle of the night
and stop all the clocks . . .

13. The Sources of Time

Perhaps God has many more seasons
in store for us –
or perhaps the last is to be
this winter
that guides back the waves
of the Tyrrhenian Sea
to break against
the rough pumice cliffs.
You must be wise. Pour the wine
and enclose in this brief circle
your long-cherished hope. (I, 11)

We started out with the image of time that is familiar to us: something that flows uniformly and equally throughout the universe, in the course of which all things happen. With the idea that there exists throughout the cosmos a present, a 'now', that constitutes reality. The past for everyone is fixed, is gone, having already happened. The future is open, yet to be determined. Reality flows from the past, through the present, towards the future – and the evolution of things between past and future is intrinsically asymmetrical. This, we thought, is the basic structure of the world.

This familiar picture has fallen apart, has shown itself to be only an approximation of a much more complex reality.

A present that is common throughout the whole universe does not exist (Chapter 3). Events are not ordered in pasts, presents and futures; they are only 'partially' ordered. There is a present that is near to us, but nothing that is 'present' in a far-off galaxy. The present is a localized rather than a global phenomenon.

The difference between past and future does not exist in the elementary equations that govern events in the world (Chapter 2). It issues only from the fact that, in the past, the world found itself subject to a state that, with our blurred take on things, appears particular to us.

Locally, time passes at different speeds according to where we are and at what speed we ourselves are moving. The closer we are to a mass (Chapter 1), or the faster we move (Chapter 3), the more time slows down: there is no single duration between two events; there are many possible ones.

The rhythms at which time flows are determined by the gravitational field, a real entity with its own dynamic that is described in the equations of Einstein. If we overlook quantum effects, time and space are aspects of a great jelly in which we are immersed (Chapter 4).

But the world *is* a quantum one, and gelatinous spacetime is also an approximation. In the elementary grammar of the world, there is neither space nor time – only processes that transform physical quantities from one to another, from which it is possible to calculate probabilities and relations (Chapter 5).

At the most fundamental level that we currently know of, therefore, there is little that resembles time as we experience it. There is no special variable 'time', there is no difference between past and future, there is no spacetime (Part Two). We still know how to write equations that describe the world. In those equations, the variables evolve with respect to each other (Chapter 8). It is not a 'static' world, or a 'block universe' where all change is illusory (Chapter 7): on the contrary, ours is a world of events rather than of things (Chapter 6).

This was the outward leg of the journey, towards a universe without time.

The return journey has been the attempt to understand how, from this world without time, it is possible for our perception of time to emerge (Chapter 9). The surprise has been that, in the emergence of familiar aspects of time, we ourselves have had a role to play. From *our perspective* – the perspective of creatures who make up a small part of the world – we see that world flowing in time. Our interaction with the world is partial, which is why we see it in a blurred way. To this blurring is added quantum indeterminacy. The ignorance that follows from this determines the existence of a particular variable – thermal time (Chapter 9) – and of an entropy that quantifies our uncertainty.

Perhaps we belong to a particular subset of the world that interacts with the rest of it in such a way that this entropy is lower in one direction of our thermal time. The directionality of time is therefore real but perspectival (Chapter 10): the entropy of the world *in relation to us* increases with our thermal time. We see the occurrence of things ordered in this variable, which we simply call 'time', and the growth of entropy distinguishes the past from the future for us and leads to the unfolding of the cosmos. It determines the existence of traces, residues and memories of the past (Chapter 11). We human beings are an effect of this great history of the increase of entropy, held together by the memory that is enabled by these traces. Each one of us is a unified being because we reflect the world, because we have formed an image of a unified entity by interacting with our kind, and because it is a perspective on the world unified by memory (Chapter 12). From this comes what we call the 'flowing' of time. This is what we are listening to when we listen to the passing of time.

The variable 'time' is one of many variables which describe the world. It is one of the variables of the gravitational field (Chapter 4): at our scale, we do not register quantum fluctuations (Chapter 5), hence it is possible to think of spacetime as determined, as Einstein's great mollusc; at our scale, the movements of the mollusc are small and can be overlooked. Hence we can think of spacetime as being as rigid as a table. This table has dimensions: the one that we call space, and the one along which entropy grows, called time. In our everyday life we move at low speeds in relation to the speed of light and so we do not perceive the discrepancies between the different proper times of different clocks, and the differences in speed at which time passes at different distances from a mass are too small for us to distinguish.

In the end, therefore, instead of many possible times we can speak only of a single time: the time of our experience: uniform, universal and ordered. This is the approximation of an approximation of an approximation of a description of the world made from our particular perspective as human beings who are dependent on the growth of entropy, anchored to the flowing of time. We for whom, as Ecclesiastes[1] has it, there is a time to be born and a time to die.

This is time for us: a multilayered, complex concept with multiple, distinct properties deriving from various different approximations.

Many discussions of the concept of time are confused because they simply do not recognize its complex and multilayered aspect. They make the mistake of not seeing that the different layers are independent.

This is the physical structure of time as I understand it, after a lifetime of revolving around it.

Many parts of this story are solid, others plausible, others still are guesses hazarded in an attempt at understanding the whole.

Practically all the things recounted in the first part

of the book have been ascertained from innumerable experiments: the slowing down of time according to altitude and speed; the non-existence of the present; the relation between time and the gravitational field; the fact that the relations between different times are dynamic, that elementary equations do not recognize the direction of time; the relation between entropy and blurring. All this has been well ascertained.[2]

That the gravitational field has quantum properties is a shared conviction, albeit one currently supported only by theoretical arguments rather than by experimental evidence.

The absence of the time variable from the fundamental equations, as discussed in Part Two, is plausible – but on the form of these equations debate still rages. The origin of time pertaining to quantum noncommutativity, of thermal time, and the fact that the increase in entropy which we observe depends on our interaction with the universe are ideas that I find fascinating but are far from being confirmed or widely accepted.

What is entirely credible, in any case, is the general fact that the temporal structure of the world is different from the naïve image that we have of it. This naïve image is suitable for our daily life, but it's not suitable for understanding the world in its minute folds or in its vastness. In all likelihood, it is not even sufficient for understanding our own nature, because the mystery of time intersects with the mystery of our personal identity, with the mystery of consciousness.

The mystery of time has always troubled us, stirring deep emotions – so deep as to have nourished philosophies and religions.

I believe, as Hans Reichenbach suggests in one of the most lucid books on the nature of time, *The Direction of Time*, that it was in order to escape from the anxiety time causes us that Parmenides wanted to deny its existence, that Plato imagined a world of ideas that exist outside of it, and that Hegel speaks of the moment in which the Spirit transcends temporality and knows itself in its plenitude. It is in order to escape this anxiety that we have imagined the existence of 'eternity', a strange world outside of time that we would like to be inhabited by gods, by a God or by immortal souls.* Our deeply emotional attitude towards time has contributed more to the

* There is something extremely interesting about the fact that this observation by Reichenbach, in a fundamental text for the treatment of time by analytical philosophy, sounds so close to ideas from which Heidegger's reflection stems. The subsequent divergence is enormous: Reichenbach searches in physics for that which we know about time in the world of which we are part, while Heidegger concerns himself with what time is in the existential experience of human beings. The two resultant images of time are completely different from each other. Are they necessarily incompatible? Why should they be? They explore two different problems: on the one hand, the effective temporal structures of the world that reveal themselves to be progressively more threadbare as we widen our gaze; on the

construction of cathedrals of philosophy than has logic or reason. The opposite emotional attitude, the veneration of time – by a Heraclitus or a Bergson – has given rise to just as many philosophies, without getting us any nearer to understanding what time is.

Physics helps us to penetrate layers of the mystery. It shows how the temporal structure of the world is different from our perception of it. It gives us the hope of being able to study the nature of time free from the fog caused by our emotions.

But in our search for time, advancing increasingly away from ourselves, we have ended up by discovering something about ourselves, perhaps – just as Copernicus, by studying the movements of the heavens, ended up by understanding how the Earth moved beneath his feet. Perhaps, ultimately, the emotional dimension of time is not the film of mist that prevents us from apprehending the nature of time objectively.

Perhaps the emotion of time is precisely what time *is* for us.

I don't think there is much more than this to be understood. We may ask further questions, but we should be careful with questions that it is not possible to formulate properly. When we have found all the aspects of time that can be spoken of, then we have found time. We may gesture clumsily towards an

---

other, the foundational aspect that the structure of time has *for us*, for our concrete sense of 'being in the world'.

immediate sense of time beyond what we can articulate ('Fine, but why does it "pass?"'), but I believe that at this point we are merely confusing matters, attempting illegitimately to transform approximate words into things. When we cannot formulate a problem with precision, it is often not because the problem is profound: it's because the problem is false.

Will we be able to understand things better in the future? I think so. Our understanding of nature has increased vertiginously over the course of centuries, and we are continuing to learn. We are glimpsing something about the mystery of time. We can see the world without time: we can perceive with the mind's eye the profound structure of the world where time as we know it no longer exists – like the Fool on the Hill who sees the Earth turn when he sees the setting sun. And we begin to see that we are time. We are this space, this clearing opened by the traces of memory inside the connections between our neurons. We are memory. We are nostalgia. We are longing for a future that will not come. The clearing that is opened up in this way, by memory and by anticipation, is time: a source of anguish sometimes, but in the end a tremendous gift.

A precious miracle that the infinite play of combinations has unlocked for us, allowing us to exist. We may smile now. We can go back to serenely immersing ourselves in time – in our finite time – to savouring the clear intensity of every fleeting and cherished moment of the brief circle of our existence.

The Sister of Sleep

> The brief arc of our days,
> O Sestius,
> prevents us from launching
> prolonged hopes. (I, 4)

In the third part of the great Indian epic the *Mahābhārata*, a powerful spirit named Yaksa asks the oldest and wisest of the Pandava, Yudhistira, what is the greatest of all mysteries. The answer given resounds across millennia: 'Every day countless people die, and yet those who remain live as if they were immortals.'[1]

I would not wish to live as if I were immortal. I do not fear death. I fear suffering. And I fear old age, though less so now that I am witnessing the tranquil and pleasant old age of my father. I am afraid of frailty, and of the absence of love. But death does not alarm me. It did not scare me when I was young, and I thought at the time that this was because it was such a remote prospect. But now, at sixty, the fear has yet to arrive. I love life, but life is also struggle, suffering, pain. I think of death as akin to a well-earned rest. The sister of sleep, Bach calls it, in his marvellous cantata

BWV 56. A kindly sister, who will come quickly to close my eyes and caress my head.

Job died when he was 'full of days'. It's a wonderful expression. I, too, would like to arrive at the point of feeling 'full of days', and to close with a smile the brief circle that is our life. I can still take pleasure in it, yes; still enjoy the moon reflected on the sea, the kisses of the woman I love, her presence that gives meaning to everything; still savour those Sunday afternoons at home in winter, lying on the sofa filling pages with symbols and formulae, dreaming of capturing another small secret from among the thousands that still surround us . . . I like to look forward to still tasting from this golden chalice, to life that is teeming, both tender and hostile, clear and inscrutable, unexpected . . . But I have already drunk deep of the bittersweet contents of this chalice, and if an angel were to come for me right now, saying, 'Carlo, it's time,' I would not ask to be left even long enough to finish this sentence. I would just smile up at him and follow.

Our fear of death seems to me to be an error of evolution. Many animals react instinctively with terror and flight at the approach of a predator. It is a healthy reaction, one which allows them to escape from danger. But it's a terror that lasts an instant, not something that remains with them constantly. Natural selection has produced these big apes with hypertrophic frontal lobes, with an exaggerated ability to predict the future. It's a prerogative that's certainly useful but one that has placed before us a vision of our inevitable death, and this triggers the instinct of terror and flight. Basically, I believe that the fear of death is the result of an accidental and clumsy interference between two distinct evolutionary pressures – the product of bad automatic connections in our brain rather than something that has any use or meaning. Everything has a limited duration, even the human race itself. ('The Earth has lost its youthfulness; it is past, like a happy dream. Now every day brings us closer to destruction, to desert . . .', as Vyasa has it, in the *Mahābhārata*.[2])Fearing the transition, being afraid of death, is like being afraid of reality itself; like being afraid of the sun. Whatever for?

This is the rational version. But our lives are not driven by rational arguments. Reason helps us to clarify ideas, to discover errors. But that same reason also shows us that the motives by which we act are inscribed in our intimate structure as mammals, as hunters, as social beings: reason illuminates these connections, it does not generate them. We are not, in the first place, reasoning beings. We may perhaps become so, more or less, in the second. In the first instance, we are driven by a thirst for life, by hunger, by the need to love, by the instinct to find our place in human society . . . The second instance does not even exist without the first. Reason arbitrates between instincts but uses the very same instincts as primary criteria in its arbitration. It gives names to things and to this thirst, it allows us to overcome obstacles, to see things that are hidden. It allows us to recognize the innumerable inefficient strategies, mistaken beliefs and prejudices that we have. It has developed to help us understand that the tracks we follow, thinking that they will lead to the antelopes we are hunting, are in fact false trails. But what drives us is not reflecting on life: it is life itself.

So what really drives us? It is difficult to say. Perhaps we do not know entirely. We recognize motivations in ourselves. We give names to these motivations, and we have many of them. We believe that we share some of them with other animals; others only with humankind – and others still with smaller groups to which we see ourselves as belonging. Hunger and thirst, curiosity, the need for companionship, the desire to love, being in love, the pursuit of happiness, the need to fight for a position in the world, the desire to be appreciated, recognized and loved; loyalty, honour, the love of God, the thirst for justice and liberty, the desire for knowledge . . .

Where does all this come from? From the way that we are made, from what we happen to be. We are the products of a long selection process of chemical, biological and cultural structures that at different levels have interacted for a long time in order to shape the funny process that we are. About which we understand very little, by reflecting on ourselves, by looking at ourselves in the mirror. We are more complex than our mental faculties are capable of grasping. The hypertrophy of our frontal lobes is considerable, and has taken us to the moon, allowed us to discover black holes and to recognize that we are cousins of ladybirds. But it is still not enough to allow us to explain ourselves clearly to ourselves.

We are not even clear about what it means 'to understand'. We see the world and we describe it: we give it an order. We know little of the actual relation between what we see of the world and the world itself. We know that we are myopic. We barely see just a tiny window of the vast electromagnetic spectrum emitted by things. We do not see the atomic structure of matter, nor the curvature of space. We see a coherent world that we extrapolate from our interaction with the universe, organized in simplistic terms that our devastatingly stupid brain is capable of handling. We think of the world in terms of stones, mountains, clouds and people, and this is 'the world for us'. About the world independent of us we know a good deal, without knowing how much this good deal is.

Our thinking is prey to its own weakness, but even more so to its own grammar. It takes only a few centuries for the world to change: from devils, angels and witches to atoms and electromagnetic waves. It takes only a few grams of mushrooms for the whole of reality to dissolve before our eyes, before reorganizing itself into a surprisingly different form. It only takes the experience of spending time with a friend who has suffered a serious schizophrenic episode, a few weeks with her struggling to communicate, to realize that delirium is a vast theatrical machinery with the capacity to stage the world, and that it is difficult to find arguments to distinguish it from those great collective deliriums of ours that are the foundations of our social and spiritual life, and of our understanding of the world. Aside, perhaps, from solitude – and the fragility of those who detach themselves from the commonplace order of things . . .[3] The vision of reality and the collective delirium that we have organized has evolved and has turned out to have worked reasonably well in getting us to this point. The instruments that we have found for dealing with it and attending to it have been many, and reason has revealed itself to be among the best of these. It is precious.

But it is only an instrument, a pincer. We use it to handle a substance that is made of fire and ice: something that we experience as living and burning emotions. These are the substances of which we are made. They propel us and they drag us back, and we cloak them with fine words. They compel us to act. And something of them always escapes from the order of our discourses, since we know that, in the end, every attempt to impose order leaves something outside the frame.

And it seems to me that life, this brief life, is nothing other than this: the incessant cry of these emotions that drive us, that we sometimes attempt to channel in the name of a god, a political faith, in a ritual that reassures us that, fundamentally, everything is in order, in a great and boundless love – and the cry is beautiful. Sometimes it is a cry of pain. Sometimes it is a song.

And song, as Augustine observed, is the awareness of time. It *is* time. It is the hymn of the Vedas that is itself the flowering of time.[4] In the Benedictus of Beethoven's *Missa Solemnis* the song of the violin is pure beauty, pure desperation, pure joy. We are suspended, holding our breath, feeling mysteriously that this must be the source of meaning. That this is the source of time.

Then the song fades and ceases. 'The silver thread is broken, the gold lantern is shattered, the amphora at the fountain breaks, the bucket falls into the well, the earth returns to dust.'[5] And it is fine like this. We can close our eyes, rest. This all seems fair and beautiful to me. This is time.

Image Credits

Pages. 10, 19, 35, 69, 77: © Peyo-2017 Licensed through I.M.P.S (Brussels) – www.smurf.com

Page 26: Ludwig Boltzmann, lithograph by Rodolf Fenzi (1899) © Hulton Archive/Getty Images

Page 54 (right-hand side): Johannes Lichtenberger, sculpture by Conrad Sifer (1493), sundial of the Cathedral of Strasbourg © Gilardi Photo Library

Page 59 (left-hand side): bust of Aristotle © De Agostini/ Getty Images

Page 59 (right-hand side): Isaac Newton, sculpture by Edward Hodges Baily (1828), after Louis-François de Roubiliac (1751), National Portrait Gallery, London © National Portrait Gallery, London/Foto Scala, Florence

Page 110: Thomas Thiemann, *Dynamic of Quantum Spin Foam, seen through the eyes of an artist* © Thomas Thiemann (FAU Erlangen), Max Planck Institute for Gravitational Physics (Albert Einstein Institute), Milde Marketing Science Communication, exozet effects

Page 135: Hildegard von Bingen, *Liber Divinorum Operum*, Codex Latinus 1942 (XIII century), c. 9r, Biblioteca Statale, Lucca © Foto Scala, Florence – by courtesy of the Ministry of Cultural Heritage and Activities

Notes

Perhaps Time is the Greatest Mystery

1. Aristotle, *Metaphysics*, I, 2, 982.
2. The layering of the notion of time is discussed in depth, for example, in J. T. Fraser, *Of Time, Passion, and Knowledge*, Braziller, New York, 1975.
3. The philosopher Mauro Dorato has insisted on the necessity to render the elementary conceptual framework of physics coherent with our experience (*Che cos'è il tempo?*, Carocci, Rome, 2013).

1. Loss of Unity

1. This is the essence of the theory of general relativity (A. Einstein, 'Die Grundlage der algemeinen Relativitätstheorie', *Annalen der Physik*, 49, 1916, pp. 769–822).
2. In the approximation of a weak field, the metrics can be written $ds^2 = \left(1 + 2\phi\left(x\right)\right)dt^2 - dx^2$ where $\phi\left(x\right)$ is the potential of Newton. Newtonian gravity follows from the sole modification of the temporal component of the metrics g_{oo}, that is, from the local slowing down of time. The geodesics of these metrics describe the fall of bodies: they bend towards the lowest potentiality, where

time slows. (These and similar notes are for those who have some familiarity with theoretical physics.)

3. Carlo Rovelli, *Che cos'è la scienza. La rivoluzione di Anassimandro*, Mondadori, Milan, 2011. English translation: *The First Scientist: Anaximander and His Legacy*, Westholme, Yardley, 2011.
4. For example: $\left(t_{table} - t_{ground}\right) = gh/c^2\, t_{ground}$ where c is the speed of light, $g = 9.8\, m/s^2$ is the acceleration of Galileo and h is the height of the table.
5. They can also be written with a single variable, t, the 'temporal coordinate', but this does not indicate the time measured by a clock (determined by ds, not by dt) and may be changed arbitrarily without changing the world described. This t does not represent a physical quantity. What clocks measure is the proper time along a line of the universe γ, given by $t_\gamma = \int_\gamma \sqrt{g_{ab}(x)\, dx^a dx^b}$. The physical relation between this quantity and ds is discussed further on.

2. *Loss of Direction*

1. Rainer Maria Rilke, *Duineser Elegien*, in *Sämtliche Werke*, Insel, Frankfurt, vol. I, 1955, I, vv. 83–5.
2. The French Revolution was an extraordinary moment of scientific vitality in which the bases of chemistry, biology, analytic mechanics and much else were founded. The social revolution went hand in hand with the scientific one. The first revolutionary mayor of Paris was an astronomer; Lazare Carnot was a mathematician; Marat considered himself to be, above all else, a physicist. Lavoisier was

active in politics. Lagrange was honoured by the different governments that succeeded each other in that tormented and magnificent moment in the history of humanity. See S. Jones, *Revolutionary Science: Transformation and Turmoil in the Age of the Guillotine*, Pegasus, New York, 2017.

3. Changing what is opportune: for instance, the sign of the magnetic field in the equations of Maxwell, charge and parity of elementary particles, etc. It is the invariance under CPT (Charge, Parity and Time reversal symmetry) that is relevant.
4. The equations of Newton determine how things *accelerate*, and the acceleration does not change if I project a film backwards. The acceleration of a stone thrown upwards is the same as that of a falling stone. If I imagine years running backwards, the moon turns around the Earth in the opposite direction but appears equally attracted to the Earth.
5. The conclusion does not change by adding quantum gravity. On the efforts to find the origin of the direction of time, see, for example, H. D. Zeh, *Die Physik der Zeitrichtung*, Springer, Berlin, 1984.
6. R. Clausius, 'Über verschiedene für die Anwendung bequeme Formen der Hauptgleichungen der mechanischen Wärmetheorie', *Annalen der Physik*, 125, 1865, pp. 353–400; p. 390.
7. In particular as a quantity of heat that escapes from a body *divided by temperature*. When the heat escapes from a hot body and enters a cold one, the total entropy increases because the difference in temperature makes it so that the entropy due to heat that escapes is less than that owed to the heat that enters. When all the bodies

reach the same temperature, the entropy has reached its maximum: equilibrium has been reached.

8. Arnold Sommerfeld.
9. Wilhelm Ostwald.
10. The definition of entropy requires a *coarse graining*, that is to say, the distinction between microstates and macrostates. The entropy of a macrostate is determined by the number of corresponding microstates. In classic thermodynamics, the *coarse graining* is defined the moment it is decided to treat some variables of the system as 'manipulable' or 'measurable' from outside (the volume or pressure of a gas, for instance). A macrostate is determined by fixing *these* macroscopic variables.
11. That is to say, in a deterministic manner if you overlook quantum mechanics, and in a probabilistic manner if you take account of quantum mechanics instead. In both cases, in the same way for the future as for the past.
12. $S = k \ln W$. Here, S is the entropy, W is the number of microscopic states, or the corresponding volume of phase space, and k is just a constant, today called Boltzmann's constant, that adjusts the (arbitrary) dimensions.

3. The End of the Present

1. General relativity (A. Einstein, 'Die Grundlage der allgemeinen Relativitätstheorie', op. cit.).
2. Special relativity (A. Einstein, 'Zur Elektrodynamik bewegter Körper', Annalen der Physik, 17, 1905, pp. 891–921).

3. J. C. Hafele and R. E. Keating, 'Around-the-World Atomic Clocks: Observed Relativistic Time Gains', Science, 177, 1972, pp. 166–8.
4. That depends as much on t as on your speed and position.
5. Poincaré. Lorentz had tried to give a physical interpretation to t, but in a quite convoluted way.
6. Einstein frequently maintained that the experiments of Michelson and Morley were of no importance in allowing him to arrive at special relativity. I believe this to be true, and that it illustrates an important factor in the philosophy of science. In order to make advances in our understanding of the world, it is not always necessary to have new data. Copernicus had no more observational data than Ptolemy: he was able to deduce heliocentrism from the data available to Ptolemy by interpreting it better – as Einstein did with regard to Maxwell.
7. If I see my sister through a telescope celebrating her twentieth birthday and send her a radio message that will arrive on her twenty-eighth birthday, I can say that *now* is her twenty-fourth birthday: halfway between when the light departed from there (20) and when it returned (28). It's a nice idea (not mine: it's Einstein's definition of 'simultaneity'). But this does not define a common time. If *Proxima b* is moving away, and my sister uses the same logic to calculate the moment simultaneous to her twenty-fourth birthday, she *does not* obtain the present moment here. In other words, in this way of defining simultaneity, if for me a moment A in her life is simultaneous with a moment B in mine, the contrary is not the case: for her, A and B are not

simultaneous. Our different speeds define different surfaces of simultaneity. Not even in this way do we obtain a notion of a common 'present'.

8. The combination of events that are at space-like distance from here.
9. Among the first to realize this was Kurt Gödel ('An Example of a New Type of Cosmological Solutions of Einstein's Field Equations of Gravitation', *Reviews of Modern Physics*, 21, 1949, pp. 447–50). In his own words: 'The notion of "now" is nothing more than a certain relation between a certain observer and the rest of the universe.'
10. Transitive.
11. Even the existence of a relation of partial order might be too strong with regard to reality, if closed temporal curves exist. On this subject see, for example, M. Lachièze-Rey, *Voyager dans le temps. La Physique moderne et la temporalité*, Éditions du Seuil, Paris, 2013.
12. The fact that there is nothing logically impossible about travels to the past is demonstrated clearly in an engaging article by one of the great philosophers of the last century, David Lewis ('The Paradoxes of Time Travel', *American Philosophical Quarterly*, 13, 1976, pp. 145–52, reprinted in *The Philosophy of Time*, eds. R. Le Poidevin and M. MacBeath, Oxford University Press, Oxford, 1993).
13. This is the representation of the causal structure of a black hole metric in Finkelstein coordinates.
14. Among the dissenting voices, there are those of two great scientists for whom I have a particular friendship, affection and admiration: Lee Smolin (*Time Reborn*, Houghton Mifflin Harcourt, Boston, 2013) and George Ellis ('On the Flow of Time', FQXi Essay, 2008, https://

arxiv.org/abs/0812.0240; 'The Evolving Block Universe and the Meshing Together of Times', *Annals of the New York Academy of Sciences*, 1326, 2014, pp. 26–41; *How Can Physics Underlie the Mind?*, Springer, Berlin, 2016). Both insist that a privileged time and a real present must exist, even if these are not captured by current physics. Science is like affection: those who are dearest to us are those with whom we have the liveliest disagreements. An articulate defence of the fundamental aspect of the reality of time can be found in R. M. Unger and Lee Smolin, *The Singular Universe and the Reality of Time* (Cambridge University Press, Cambridge, 2015). Another dear friend who defends the idea of the real flowing of a singular time is Samy Maroun; with him I have explored the possibility of rewriting the physics of relativity, distinguishing the time that guides the rhythm of processes ('metabolic' time) from a 'real' universal time (S. Maroun and C. Rovelli, 'Universal Time and Spacetime "Metabolism"', 2015). This is possible, and hence the point of view of Smolin, Ellis and Maroun is defensible. But is it fruitful? The choice is between forcing the description of the world so that it adapts to our intuition, or learning instead to adapt our intuition to what we have discovered about the world. I have few doubts that the second strategy is the most fruitful one.

4. Loss of Independence

1. On the effects of drugs on time perception, see R. A. Sewell et al., 'Acute Effects of THC on Time Perception

in Frequent and Infrequent Cannabis Users', *Psychopharmacology*, 226, 2013, pp. 401–13; the direct experience is astonishing.

2. V. Arstila, 'Time Slows Down during Accidents', *Frontiers in Psychology*, 3, 196, 2012.
3. In our cultures. There are others with a profoundly different notion of time: D. L. Everett, *Don't Sleep, There are Snakes*, Pantheon, New York, 2008.
4. Matthew 20:1–16.
5. P. Galison, *Einstein's Clocks, Poincaré's Maps*, Norton, New York, 2003, p. 126.
6. An excellent panoramic history of the way in which technology has progressively modified our concept of time can be found in A. Frank, *About Time*, Free Press, New York, 2001.
7. D. A. Golombek, I. L. Bussi and P. V. Agostino, 'Minutes, Days and Years: Molecular Interactions among Different Scales of Biological Timing', *Philosophical Transactions of the Royal Society. Series B: Biological Sciences*, 369, 2014.
8. Time is: 'number of change, with regard to before and after' (Aristotle, *Physics*, IV, 219 b 2; see also 232 b 22–3).
9. Aristotle, *Physics*, trans. Robin Waterfield with an introduction and notes by David Bostock, Oxford University Press, Oxford, 1999, p. 105.
10. Isaac Newton, *Philosophiae Naturalis Principia Mathematica*, Book I, def. VIII, scholium.
11. Ibid.
12. An introduction to the philosophy of space and of time can be found in B. C. van Fraassen, *An Introduction to the Philosophy of Time and Space*, Random House, New York, 1970.

13. Newton's fundamental equation is $F = m\, d^2x/dt^2$. (Note that time t is squared: this indicates that the equation does not distinguish t from $-t$, that is to say, it is the same backwards or forwards in time, as I explain in Chapter 2.
14. Curiously, many contemporary manuals of the history of science present the discussion between Leibniz and the Newtonians as if Leibniz were the heterodox figure with audacious and innovative relationist ideas. In reality, the opposite was the case: Leibniz defended (with a new wealth of arguments) the dominant traditional understanding of space, which from Aristotle to Descartes had always been relationist.
15. Aristotle's definition is more precise: the place of a thing is the *inner boundary* of that which surrounds the thing. An elegant and rigorous definition.

5. Quanta of Time

1. I speak of this in more depth in *Reality is Not What It Seems*, trans. Simon Carnell and Erica Segre, Allen Lane, London, 2016.
2. It is not possible to locate a degree of liberty in a region of its phase space within a volume smaller than the Planck constant.
3. The speed of light, the Newton constant and the Planck constant.
4. Maimonides, *The Guide for the Perplexed*, I, 73, 106a.
5. We can try to infer the thought of Democritus from the discussions of Aristotle (for example, in *Physics*, IV, 213), but the evidence seems insufficient to me. See *Democrito.*

Raccolta dei frammenti, interpretazione e commentario di Salomon Luria, Bompiani, Milan, 2007.

6. Unless the de Broglie–Bohm theory is true, in which case it has it – but hides it from us. Which is perhaps not so different in the end.
7. Carlo Rovelli, 'Relational Quantum Mechanics', *International Journal of Theoretical Physics*, 35, 1637 (1996), http://arxiv.org/abs/quant-ph/9609002. See also 'The Sky is Blue and Birds Fly Through It', http://arxiv.org/abs/1712.02894.
8. Grateful Dead, 'Walk in the Sunshine'.

6. The World is Made of Events, not Things

1. Nelson Goodman, *The Structure of Appearance*, Harvard University Press, Cambridge, MA, 1951.

7. The Inadequacy of Grammar

1. For opposing views, see Chapter 3, note 14.
2. In the terminology of a celebrated article by John McTaggart ('The Unreality of Time', *Mind*, N.S., 17, 1908, pp. 457–74; reprinted in *The Philosophy of Time*, op. cit.), this is equivalent to denying the reality of the A-series (the organization of time into 'past–present–future'). The meaning of temporal determinations would then be reduced to only the B-series (the organization of time into 'before-it, after-it'). For McTaggart, this implies denying the reality of time. To my mind, McTaggart is too inflexible: the fact that my car works differently from

how I'd imagined it and how I'd originally defined it in my head does not mean that my car is not real.

3. Letter by Einstein to the son and sister of Michele Besso, 21 March 1955, in Albert Einstein and Michele Besso, *Correspondance, 1903–1955*, Hermann, Paris, 1972.
4. The classic argument for the block universe is given by the philosopher Hilary Putnam in a famous article published in 1967 ('Time and Physical Geometry', *Journal of Philosophy*, 64, pp. 240–47). Putnam uses Einstein's definition of simultaneity. As we have seen in Chapter 3, note 7, if the Earth and *Proxima b* move with respect to one another, say they are approaching each other, an event A on Earth is simultaneous (for an earthling) to an event B on *Proxima b*, which in turn is simultaneous (for those on *Proxima b*) to an event C on Earth, *that is in the future of A*. Putnam assumes that 'being simultaneous' implies 'being real now', and deduces that the event in the future (such as C) is real now. The error is to assume that Einstein's definition of simultaneity has an ontological value, whereas it is only a definition of convenience. It serves to identify a relativistic notion that may be reduced to the non-relativistic one through an approximation. But non-relativistic simultaneity is a notion which is reflexive and transitive, whereas Einstein's is not, hence it makes no sense to assume that the two have the same ontological meaning beyond the approximation.
5. That the discovery by physics of the impossibility of presentism implies that time is illusory is an argument put forward by Gödel ('A Remark about the Relationship between Relativity Theory and Idealistic Philosophy', in *Albert Einstein: Philosopher-Scientist*, ed. P. A. Schlipp, Library of Living Philosophers, Evanston, 1949). The

error always lies in defining time as a single conceptual block that is either all there or not there at all. The point is discussed lucidly by Mauro Dorato (*Che cos'è il tempo?*, op. cit., p. 77).

6. See, for instance, W. V. O. Quine, 'On What There Is', *Review of Metaphysics*, 2, 1948, pp. 21–38, and the fine discussion of the meaning of reality in J. L. Austin, *Sense and Sensibilia*, Clarendon Press, Oxford, 1962.
7. *De Hebd.*, II, 24, cited in C. H. Kahn, *Anaximander and the Origins of Greek Cosmology*, Columbia University Press, New York, 1960, pp. 84–5.
8. Some examples of important arguments where Einstein has strongly supported a thesis which he later changed his mind about: 1. The expansion of the universe (first ridiculed, then accepted); 2. the existence of gravitational waves (first taken as obvious, then rejected, then accepted again); 3. the equations of relativity do not admit solutions without matter (a long-defended thesis that was abandoned – rightly so); 4. nothing exists beyond the horizon of Schwarzschild (wrong, though perhaps he never came to realize this); 5. the equations of the gravitational field cannot be general covariant (asserted in the work with Grossmann in 1912; three years later, Einstein argued the opposite); 6. the importance of the cosmological constant (first affirmed, then denied – having been right the first time) . . .

8. Dynamics as Relation

1. The general form of a mechanical theory that describes the development of a system *in time* is given by a phase

space and a Hamiltonian H. Evolution is described by the orbits generated by H, parametrized by the time t. The general form of a mechanical theory that describes the evolutions of variables *with respect to each other* is instead given by a phase space and a constraint C. The relations between the variables are given by the orbits generated by C in the subspace $C = 0$. The parametrization of these orbits has no physical meaning. A detailed technical discussion can be found in Chapter 3 of Carlo Rovelli, *Quantum Gravity*, Cambridge University Press, Cambridge, 2004. For a concise technical account, see Carlo Rovelli, 'Forget Time', *Foundations of Physics*, 41, 2011, pp. 1475–90, https://arxiv.org/abs/0903.3832.

2. An accessible account of loop quantum gravity can be found in Rovelli, *Reality is Not What It Seems*, op. cit.
3. B. S. DeWitt, 'Quantum Theory of Gravity. I. The Canonical Theory', *Physical Review*, 160, 1967, pp. 1113–48.
4. J. A. Wheeler, 'Hermann Weyl and the Unity of Knowledge', *American Scientist*, 74, 1986, pp. 366–75.
5. J. Butterfield and C. J. Isham, 'On the Emergence of Time in Quantum Gravity', in *The Arguments of Time*, ed. J. Butterfield, Oxford University Press, Oxford, 1999, pp. 111–68 (http://philsci-archive.pitt.edu/1914/1/EmergTimeQG=9901024.pdf). H.-D. Zeh, *Die Physik der Zeitrichtung*, op. cit., *Physics Meets Philosophy at the Planck Scale*, ed. C. Callender and N. Huggett, Cambridge University Press, Cambridge, 2001. S. Carroll, *From Eternity to Here*, Dutton, New York, 2010.
6. The general form of a quantum theory that describes the evolution of a system *in time* is given by a Hilbert space and the Hamiltonian operator H. The evolution is described

by Schrödinger's equation $i\hbar\partial_t\Psi = H\Psi$. The probability of measuring a pure state Ψ a time t after having measured a state Ψ' is given by the transition amplitude $\langle\Psi|\exp[-iHt/\hbar]|\Psi'\rangle$. The general form of a quantum theory that describes the evolution of the variables *with respect to one another* is given by a Hilbert space and Wheeler–DeWitt equation $C\Psi = 0$. The probability of measuring the state Ψ' after having measured the state Ψ is determined by the amplitude $\left\langle\Psi\middle|\int dt \exp[-iCt/\hbar]\middle|\Psi'\right\rangle$. A detailed technical discussion can be found in Chapter 5 of Carlo Rovelli, *Quantum Gravity*, op. cit. For a concise technical version, see Carlo Rovelli, 'Forget Time', op. cit.

7. B. S. DeWitt, *Sopra un raggio di luce*, Di Renzo, Rome, 2005.
8. There are three: they define the Hilbert space of the theory where the elementary operators are defined, whose eigenstates describe the quanta of space and the probability of transitions between these.
9. Spin is the quantity that enumerates the representations of the group SO(3), the group of spatial symmetry. The mathematics that describes the spin networks has this feature in common with the mathematics of ordinary physical space.
10. These arguments are covered in detail in Carlo Rovelli, *Reality is Not What It Seems*, op. cit.

9. Time is Ignorance

1. Cf. Ecclesiastes 3: 2–4.
2. More precisely, the Hamiltonian H, that is, the energy as a function of position and speed.

3. $dA / dt = \{A, H\}$, where $\{ , \}$ are the Poisson brackets and A is any variable.
4. Ergodic.
5. The equations are more readable in the canonical formations of Boltzmann than in the microcanonical form to which I make reference in the text: the state $\rho = \exp[-H / kT]$ is determined by the Hamiltonian H that generates evolution of time.
6. $H = -kT \ln[\rho]$ determines a Hamiltonian (up to a multiplicative constant), and with this a 'thermal' time, starting from the state ρ.
7. Roger Penrose, *The Emperor's New Mind*, Oxford University Press, Oxford, 1989; *The Road to Reality*, Cape, London, 2004.
8. In the language of quantum mechanics manuals, it is conventionally referred to as 'measure'. Once again, there is something misleading about this language, inasmuch as it speaks about physics laboratories rather than speaking about the world.
9. The theorem of Tomita–Takesaki shows that a state on a von Neumann algebra defines a flow (a one-parameter family of modular automorphisms). Connes has shown that the flows defined by different states are equivalent up to internal automorphisms, and therefore define an abstract flow determined *only* by the noncommutative structure of the algebra.
10. The internal automorphisms of the algebra referred to in the above note.
11. In a von Neumann algebra the thermal time of a state is exactly the same as Tomita's flow! The state is KMS with respect to this flow.

12. See Carlo Rovelli, 'Statistical Mechanics of Gravity and the Thermodynamical Origin of Time', *Classical and Quantum Gravity*, 10, 1993, pp. 1549–66; Alain Connes and Carlo Rovelli, 'Von Neumann Algebra Automorphisms and Time–Thermodynamics Relation in General Covariant Quantum Theories', *Classical and Quantum Gravity*, 11, 1994, pp. 2899–918.
13. A. Connes, D. Chéreau and H. Dixmier, *Le Théâtre quantique*, Odile Jacob, Paris, 2013.

10. Perspective

1. There are many confused aspects to this question. An excellent, cogent critique can be found in J. Earman, 'The "Past Hypothesis": Not Even False', *Studies in History and Philosophy of Modern Physics*, 37, 2006, pp. 399–430. In the text, 'low initial entropy' is intended in the more general sense that, as Earman argues in this article, is far from being well understood.
2. Friedrich Nietzsche, *The Gay Science*, trans. with commentary by Walter Kaufman, Vintage, New York, 1974, p. 297.
3. The technical details can be found in Carlo Rovelli, 'Is Time's Arrow Perspectival?' (2015), in *The Philosophy of Cosmology*, ed. K. Chamcham, J. Silk, J. D. Barrow and S. Saunders, Cambridge University Press, Cambridge, 2017, https://arxiv.org/abs/1505.01125.
4. In the classical formulation of thermodynamics we describe a system by specifying in the first place some variables on which we assume we can act from the outside (moving a piston, for example), or which we assume we can measure (a

relative concentration of components, for example). These are the 'thermodynamic variables'. Thermodynamics is not a true description of the system, it is a description of *these* variables of the system: those through which we assume we are able to interact with the system.

5. For example, the entropy of the air in this room has a value taken from air as a homogeneous gas, but it changes (diminishes) if I measure its chemical composition.
6. A contemporary philosopher who has shed light on these aspects of the perspectival nature of the world is Jenann T. Ismael, in *The Situated Self* (Oxford University Press, New York, 2007). Ismael has also written an excellent book on free will: *How Physics Makes Us Free* (Oxford University Press, New York, 2016).
7. David Z. Albert (*Time and Chance*, Harvard University Press, Cambridge, MA, 2000) proposes to elevate this fact to a natural law, and calls it the *past hypothesis*.

11. What Emerges from a Particularity

1. This is another common source of confusion, because a condensed cloud seems more 'ordered' than a dispersed one. It isn't, because the speed of the molecules of a dispersed cloud are all small (in an ordered manner), while, when the cloud condenses, the speeds of the molecules increase and spread in phase space. The molecules concentrate in physical space but disperse in phase space, which is the relevant one.
2. See, in particular, S. A. Kauffman, *Humanity in a Creative Universe*, Oxford University Press, New York, 2016.

3. The importance of the existence of this ramified structure of interactions in the universe for the understanding of the growth of local entropy is discussed, for instance, by Hans Reichenbach (*The Direction of Time*, University of California Press, Berkeley, 1956). Reichenbach's text is fundamental for whoever has doubts about these arguments or is interested in pursuing them in more depth.
4. On the precise relation between traces and entropy, see Hans Reichenbach, *The Direction of Time*, op. cit., in particular, the discussion on the relation between entropy, traces and *common cause*, and D. Z. Albert, *Time and Chance*, op. cit. A recent approach can be found in D. H. Wolpert, 'Memory Systems, Computation and the Second Law of Thermodynamics', *International Journal of Theoretical Physics*, 31, 1992, pp. 743–85.
5. On the difficult question of what 'cause' means to us, see N. Cartwright, *Hunting Causes and Using Them*, Cambridge University Press, New York, 2007.
6. 'Common cause', in Reichenbach's terminology.
7. Bertrand Russell, 'On the Notion of Cause', *Proceedings of the Aristotelian Society*, N. S., 13, 1912–1913, pp. 1–26.
8. N. Cartwright, *Hunting Causes and Using Them*, op. cit.
9. For a lucid discussion on the question of the direction of time, see H. Price, *Time's Arrow and Archimedes' Point*, Oxford University Press, Oxford, 1996.

12. The Scent of the Madeleine

1. *Mil.*, II, 1, in *Sacred Books of the East*, vol. XXXV, 1890.

2. Carlo Rovelli, *Meaning = Information + Evolution*, 2016, https://arxiv.org/abs/1611.02420.
3. G. Tononi, O. Sporns and G. M. Edelman, 'A Measure for Brain Complexity: Relating Functional Segregation and Integration in the Nervous System', *Proceedings of the National Academy of Sciences USA*, 91, 1994, pp. 5033–7.
4. J. Hohwy, *The Predictive Mind*, Oxford University Press, Oxford, 2013.
5. See, for example, V. Mante, D. Sussillo, K. V. Shenoy and W. T. Newsome, 'Context-dependent Computation by Recurrent Dynamics in the Prefrontal Cortex', *Nature*, 503, 2013, pp. 78–84, and the literature cited in this article.
6. D. Buonomano, *Your Brain is a Time Machine: The Neuroscience and Physics of Time*, Norton, New York, 2017.
7. *La Condemnation parisienne de 1277*, ed. D. Piché, Vrin, Paris, 1999.
8. Edmund Husserl, *Vorlesungen zur Phänomenologie des inneren Zeitbewusstseins*, Niemeyer, Halle a. d. Saale, 1928.
9. In the cited text, Husserl insists that this does not constitute a 'physical phenomenon'. To a naturalist, this sounds like a statement of principle: he does not *want* to see memory as a physical phenomenon because he has *decided* to use phenomenological experience as the starting point of his analysis. The study of the dynamics of neurons in our brain shows how the phenomenon manifests itself in physical terms: the present of the physical state of my brain 'retains' its past state, and this is gradually more faded the further away we are from that past. See, for example, M. Jazayeri and M. N. Shadlen, 'A Neural Mechanism for Sensing and Reproducing a Time Interval', *Current Biology*, 25, 2015, pp. 2599–609.

10. Martin Heidegger, 'Einführung in die Metaphysik' (1935), in *Gesamtausgabe*, Klostermann, Frankfurt am Main, vol. XL, 1983, p. 90.
11. Martin Heidegger, *Sein und Zeit* (1927), in *Gesamtausgabe*, op. cit., vol. II, 1977, *passim*; trans. as *Being and Time*.
12. Marcel Proust, *Du côté du chez Swann*, in *À la Recherche du temps perdu*, Gallimard, Paris, vol. I, 1987, pp. 3–9.
13. Ibid., p. 182.
14. G. B. Vicario, *Il tempo. Saggio di psicologia sperimentale*, Il Mulino, Bologna, 2005.
15. The observation, a quite common one, can be found, for example, in the introduction to J. M. E. McTaggart, *The Nature of Existence*, Cambridge University Press, Cambridge, vol. I, 1921.
16. *Lichtung*, perhaps, in Martin Heidegger, *Holzwege* (1950), in *Gesamtausgabe*, op. cit., vol. V, 1977, *passim*.
17. For Durkheim (*Les Formes élémentaires de la vie religieuse*, Alcan, Paris, 1912), one of the founders of sociology, like the other great categories of thought, the concept of time has its origins in society – and in particular in the religious structure that constitutes its primary form. If this can be true for complex aspects of the notion of time – for the 'more external layers' of the notion of time – it seems to me difficult to extend it to include our direct experience of the passage of time: other mammals have brains roughly similar to ours, and consequently *experience* the passage of time like we do, without any need for a society or a religion.
18. On the foundational aspect of time for human psychology, see William James's classic *The Principles of Psychology*, Henry Holt, New York, 1890.

19. *Mahāvagga*, I, 6, 19, in *Sacred Books of the East*, vol. XIII, 1881. For the concepts relating to Buddhism, I have drawn particularly on H. Oldenburg, *Buddha*, Dall'Oglio, Milan, 1956.
20. Hugo von Hofmannstahl, *Der Rosenkavalier*, Act I.

13. The Sources of Time

1. Ecclesiastes 3:2.
2. For a light-hearted, engaging but informed exposition of these aspects of time, see C. Callender and R. Edney, *Introducing Time*, Icon Books, Cambridge, 2001.

The Sister of Sleep

1. *Mbh*, III, 297.
2. Cf. *Mbh*, I, 119.
3. A. Balestrieri, 'Il disturbo schizofrenico nell'evoluzione della mente umana. Pensiero astratto e perdita del senso naturale della realtà', *Comprendre*, 14, 2004, pp. 55–60.
4. Roberto Calasso, *L'ardore*, Adelphi, Milan, 2010.
5. Ecclesiastes 12:6–7.

Index

acceleration 16n, 186n4, 187n4
Albert, David Z. 201n7
Anaxandridas 42, 43
Anaximander 10, 13–14, 89
Aristotle 57–9, 62–4, 69, 70, 75, 86, 193n5, 193nn14–15
astronomy 14, 59, 90
atomism 76
atoms 27, 76, 88, 90–91, 180
Augustine of Hippo 156–8, 182

Bach, Johann Sebastian: BWV 56 cantata 176–7
Bede, the Venerable 76
Beethoven, Ludwig van: *Missa Solemnis* 182
Besso, Michele 101
black holes 49–50, 110–11
block universe 96–7, 195n4
Boltzmann, Ludwig 25–33, 120, 126, 137, 199n5
Boltzmann's constant 188n12
brain 134, 144, 145, 152, 155, 156, 157, 161–2, 164, 177–8, 180
  and fear of death 177–8
  frontal lobes 177, 180
  neurons *see* neurons/neural structures
  and Proust 163
  synapses 162, 163
Buddha 164

Carnot, Lazare 20–21, 186n2
Carnot, (Nicolas Léonard) Sadi 21–2, 26
cause and effect 19, 20, 30, 146–7, 152
change
  and complexity 97
  relations, events and world dynamics without a time variable 102–12, 169
  time as measurement of change, and Aristotle's space 57–71, 86
  time when nothing changes 57–8, 65
  and world as network of events 85–92, 169
Clausius, Rudolf 22, 23–5, 137
  equation of entropy 24–6, 137, 163
Cleomenes I 41–2, 43
clocks 14–15, 53–6, 60, 67, 86, 171, 186n5

clocks – (*cont.*)
and effect of speed on time 36–7n, 171
slowed by a mass/gravitational field 9–13, 68–9
synchronization of 55–6
Connes, Alain 121–4, 199n9
consciousness 56, 149, 158–62, 172–5
internal consciousness of time 159, 161
continuity 75, 85
Copernicus, Nicolaus 10–11, 174, 189n6

death 176
fear of 176, 177–8
delirium, collective 181
Democritus 76, 90, 193n5
Descartes, René 153, 193n14
DeWitt, Bryce 104, 105–6, 107
diurnal rhythms 56
Dorato, Mauro 185n3
Duino 33
Durkheim, Émile 204n17

Earman, J. 200n1
Ecclesiastes 115, 171
Einstein, Albert 10, 11–12, 34–6, 40, 47–8, 170, 189n6, 196n8
and Besso 101
general theory of relativity 15–16
and the gravitational field 66–71, 168
on illusion of time 96, 100–101
simultaneity 189–90n7, 195n4
special relativity theory 43–4, 189n6
at Swiss Patent Office 56
electrons 75, 77, 79–80, 108, 121
Ellis, George 190–91n14
energy 138–9
conservation of 25, 118–19, 138
thermal *see* heat
and thermal time 118–20
see also thermal time
entropy 23–33, 120, 125, 126–31, 136–7, 138–44, 169–70, 171, 187–8n7, 188n12, 201n5
and blurring 20–34, 118–20, 126–31, 134, 136–7, 172
Clausius's equation 24–6, 137, 163
and coarse graining 188n10
in distant past 125, 128–31, 140, 141–2, 145, 147
increase 140–44, 163–4, 170, 171, 187–8n7
low 28–9, 125, 128–31, 136–7, 139–41, 145, 147, 170
and perspective 128–37
second law of thermodynamics 22–7, 130, 139, 143

eternalism 96–7
  see also block universe
events
  relations, events and loop theory without a time variable 102–12
  world as network of 85–92, 169
evolution 177–80

fear of death 176, 177–8
flow of time 1, 2–3, 167, 170
  and cause and effect 19, 20, 30, 146–7
  Durkheim, religion and 204n17
  heat, entropy and our blurred vision of 20–34, 118–20, 126–31, 134, 136–7, 172
  in Hindu mythology 1
  internal consciousness of 159, 161
  and particularity 15, 28–32, 81, 128–31, 140–41, 147, 163–4, 168, 169–70, 171
  past, future, and the arrow of time 18–34, 128–31, 168, 169, 170
  and relativity 121 *see also* relativity
  slowed by a mass 9–13, 68–9, 168, 171, 172
  slowed by speed 34–7, 168, 171, 172
  source of Rilke's eternal current 19–34
  and the structure of spacetime 43–51, 66–71, 94–5, 97, 159–60
  and subjectivity/identity/consciousness 2–3, 4, 19–20, 52–3, 57, 148–66, 172–5 *see also* perspective
French Revolution 20, 186–7n2

Gödel, Kurt 48–9, 190n9, 195n5
Gorgo, queen of Sparta 41–2, 43
grammar
  of change 86–92
  inadequacy of 93–101, 115–17
granularity, quantum mechanics 74–6, 108
  spin networks 109–10, 111, 198n9
gravitational field 66–71, 80, 108, 168, 170, 172
  and granularity 74, 108–9

Hamiltonian H 197n1, 197n6, 199nn5–6
heat 20–24, 139, 143
  and entropy *see* entropy
  thermal time 118–21, 122–3, 124, 169, 170, 199n11
  thermodynamics *see* thermodynamics

Hegel, Georg Wilhelm Friedrich 173
Heidegger, Martin 75, 161, 173n
helium 141, 142
Hilbert space 197–8n6, 198n8
Hildegard of Bingen 134–5
Hindu mythology 1, 17, 144, 176, 178
Hofmannstahl, Hugo von: *Der Rosenkavalier* 165–6
Hume, David 152
Husserl, Edmund 159, 160–61, 203n9
hydrogen 141, 142

identity 148–66, 172
impermanence 87
  see also change
indeterminacy, quantum mechanics 77–9, 108–12, 123, 169
  and blurring 123–4, 147
indexicality 132–6
Isham, Chris 105–6
Isidore of Seville 75–6
Ismael, Jenann T. 201n6

Job 177

Kant, Immanuel 159–60
Kepler, Johannes 90, 91
Kuchar, Karel 105–6

Lagrange, Joseph-Louis 187n2
Lavoisier, Antoine 186–7n2
Leibniz, Gottfried Wilhelm 61, 193n14
Leonidas I 41–2, 43
light 38, 45, 66
  cones 45–50, 69, 78
  granularity of 75, 108
  and indeterminacy in quantum mechanics 78
  photons 75, 108, 139–40
loop theory/loop quantum gravity 72–3, 75, 107–12
Lorentz, Hendrik 189n5
Louis XVI 20

Mahābhārata 176, 178
Maimonides 76
Marat, Jean-Paul 186n2
Maroun, Samy 191n14
mass, time slowed by 9–13, 68–9, 168, 171, 172
Matthew, Gospel of 53
Maxwell's equations 14, 35–6, 91–2, 187n3
McTaggart, John 194n2
memory/memories 20, 105, 130, 144, 147, 170
  identity, time and 154–66
Milinda Pañha 148–50
music 157–8, 160, 165–6, 182

nervous system 151–2, 155–6, 162
neurons/neural structures 56, 151–2, 155–6, 163, 175, 203n9
Newton, Isaac 58–61, 62, 63, 64–8, 69–70, 102
  equations 14, 60, 187n4, 193n13
  mechanics 14, 91–2
  Philosophiae Naturalis Principia Mathematica 57–8
Newtonian gravity 185n2
noncommutativity
  quantum 121–3, 172
  Von Neumann algebra 122, 199n9, 199n11
nostalgia 105, 107, 175
nuclear fusion 141–2

Ockham, William of 158–9

Parmenides 173
partial order 42–4, 122, 190n11
particularity 15, 28–32, 81, 128–31, 140–41, 147, 163–4, 168, 169–70, 171
past traces 144–5, 157, 158, 162, 170
  see also memory/memories
Penrose, Roger 121
perspective 3, 4, 106, 125–37, 169–70
  blurring 20–34, 118–20, 123–4, 126–31, 134, 147, 169, 172
  and entropy 128–37 *see also* entropy: and blurring
  and indexicality 132–6
  and particularity *see* particularity
  subjectivity/identity/consciousness and time 2–3, 4, 19–20, 52–3, 57, 148–66, 172–5
photons 75, 108, 139–40
photosynthesis 142
Planck constant 193nn2–3
Planck length 76
Planck scale 74, 109
Planck time 74
Plato 90, 91, 173
Pleistarchus 41–2, 43
prediction 151–2, 155, 163–4, 177
  and loop theory 75, 112
present 30, 40, 167
  meaninglessness of 'now', and loss of the present 34–51, 95–6, 168
  temporal structure of universe without the present 41–51, 94–5, 97
presentism 93–4, 95–6, 195–6n5
probability, quantum mechanics *see* indeterminacy, quantum mechanics
proper time 15, 37, 186n5
Proust, Marcel 162–3
Putnam, Hilary 195n4

quantum field theory 14
quantum gravity 4, 72, 80, 103–4
  loop quantum gravity 72–3, 75, 107–12
quantum mechanics 14, 71, 72–81, 92, 168–9
  and granularity 73, 74–6, 108
  and indeterminacy *see* indeterminacy, quantum mechanics
  and loop quantum gravity 72–3, 75, 107–12
  noncommutativity 121–3, 172
  and quanta of time 72–81 *see also* quantum time
  quantum superpositions of times 77–8
  relationality 79–80
  and relativity 72–4, 121
quantum time 121–4
  spinfoam and 'quantum spacetime' 110

reality 1–2, 66–7, 69–70, 88
  and the block universe 96–7, 195n4
  collective delirium and our vision of 181
  and inadequacy of grammar 93–101
  and memory 154–66 *see also* memory/memories
  relations, events and world dynamics without a time variable 102–12, 169
reason 178–9
Reichenbach, Hans 173, 202n3, 202n4
relationality
  quantum mechanics 79–80
  relations, events and world dynamics without a time variable 102–12, 169
relativity 13–17, 42, 127
  and experience of flow of time 121
  general theory of 15–16
  and quantum mechanics 72–4, 121 *see also* quantum mechanics
  special 43–4, 189n6
Rilke, Rainer Maria: *Duino Elegies* 19, 33
Russell, Bertrand 146–7

Saadi Shirazi 20–21
Schrödinger equations 14, 91, 197–8n6
Shiva 1, 17, 144
simultaneity 189–90n7, 195n4
slowing of time 9–13, 34–7, 68–9, 168, 171, 172
Smolin, Lee 190–91n14
Sophocles: *Oedipus Rex* 5

spacetime
  and Aristotle 57–71, 86
  curved 69, 180
  'empty space' 63–5
  and eternalism 96–7 *see also* block universe
  fluctuations 77–8
  granularity 74–6, 108
  gravitational field *see* gravitational field
  loss of 85–112, 169
  and quantum mechanics *see* quantum mechanics
  spinfoam and 'quantum spacetime' 110 *see also* loop theory/loop quantum gravity
  structure of 43–51, 66–71, 94–5, 97, 159–60, 170–71
  'true' time and Newton's space 58–71
special relativity theory 43–4, 189n6
speed 127
  acceleration 16n, 186n4, 187n4
  time slowed by 34–7, 168, 171
spin networks 109–10, 111, 198n9
spinfoam 110
Strasburg Cathedral 54
Strauss, Richard: *Der Rosenkavalier* 165–6
string theory 73
suffering 101, 164–5, 176
sundials 53, 54–5
synapses 162, 163

Tempier, Étienne 158
temporal precedence 43–4
thermal time 118–21, 122–3, 124, 169, 170, 199n11
thermodynamics
  second law of 22–7, 130, 139, 143
  and thermal time 120, 124 *see also* thermal time
  variables 200–201n4
time
  absence, and equations without a time variable 86, 102–12, 169, 172, 193n13
  Anaximander and the order of 13–14, 89
  arrow of 18–34, 128–31, 168, 169, 170
  Augustine on 156–8
  clock-regulated 53–6 *see also* clocks
  closed lines 48n
  and diurnal rhythms 56
  dividing and measuring 53–7
  Einstein on illusion of 96, 100–101
  and emotion 124, 173–4, 181–2
  and eternalism 96–7 *see also* block universe
  flow of *see* flow of time

time – (*cont.*)
and grammar *see* grammar
granularity of 75–6
and the gravitational field 66–71, 168, 170, 172 *see also* gravitational field
and heat *see* heat
internal consciousness of 159, 161
loss of direction 18–34, 168, 169
loss of independence 52–71
loss of the present 34–51, 95–6, 168
loss of unity 9–17
as measurement of change, and Aristotle's space 57–71, 86
and memory *see* memory/memories
modification of structure of 9–13
multilayered aspect of 3, 171
multiple meanings of 16n
mystery of 1–5, 172–5
and network of relations 79–80 *see also* relationality
when nothing changes 57–8, 65
Planck time 74
proper time 15, 37, 186n5
quantum superpositions of times 77–8 *see also* quantum mechanics
quantum time *see* quantum time
and relativity *see* relativity
sources of 167–75
spacetime *see* spacetime
standardized 55–6
as suffering 164–5
temporal structure of universe without the present 41–51, 94–5, 97
thermal 118–21, 122–3, 124, 169, 170, 199n11
traces of the past 144–5, 157, 158, 162, 170 *see also* memory/memories
'true' time and Newton's space 58–71
Tomita–Tăkesaki theorem 199n9

universal man 135

Wheeler, John 104–5, 106, 107
Wheeler-DeWitt equation 104, 198n6